SpringerBriefs in Applied Sciences and Technology

Computational Intelligence

Series editor

Janusz Kacprzyk, Polish Academy of Sciences, Systems Research Institute, Warsaw, Poland

The series "Studies in Computational Intelligence" (SCI) publishes new developments and advances in the various areas of computational intelligence—quickly and with a high quality. The intent is to cover the theory, applications, and design methods of computational intelligence, as embedded in the fields of engineering, computer science, physics and life sciences, as well as the methodologies behind them. The series contains monographs, lecture notes and edited volumes in computational intelligence spanning the areas of neural networks, connectionist systems, genetic algorithms, evolutionary computation, artificial intelligence, cellular automata, self-organizing systems, soft computing, fuzzy systems, and hybrid intelligent systems. Of particular value to both the contributors and the readership are the short publication timeframe and the world-wide distribution, which enable both wide and rapid dissemination of research output.

More information about this series at http://www.springer.com/series/10618

Noorhazlinda Abd Rahman

Crowd Behavior Simulation of Pedestrians During Evacuation Process

DEM-Based Approach

 Springer

Noorhazlinda Abd Rahman
School of Civil Engineering
Universiti Sains Malaysia
Nibong Tebal, Penang, Malaysia

ISSN 2191-530X   ISSN 2191-5318   (electronic)
SpringerBriefs in Applied Sciences and Technology
ISSN 2625-3704   ISSN 2625-3712   (electronic)
SpringerBriefs in Computational Intelligence
ISBN 978-981-13-1845-0   ISBN 978-981-13-1846-7   (eBook)
https://doi.org/10.1007/978-981-13-1846-7

Library of Congress Control Number: 2018949870

This Springer imprint is published by the registered company Springer Nature Singapore Pte Ltd.
The registered company address is: 152 Beach Road, #21-01/04 Gateway East, Singapore 189721, Singapore

Preface

The motivation to understand crowd behavior initiates by the desire to unveil the essential laws of the variety of complex collective behaviors emerging from individual interaction in a crowd. And, this desire is growing rapidly until now to the stage of the demand for important applications in designing and planning of crowd safety.

To date, several established models for crowd evacuation have been proposed to uncover the vital laws of evacuation dynamics. Interest in crowd evacuation modeling research has been performed since 1970s. It originates by considering the crowd as a continuous flow of the medium in which the kinetic equations describing the evacuation process of the entire crowd were established. The characteristics of crowd motion are described macroscopically without the consideration of behavioral characteristics of the individual in the crowd. The evacuation studies have grown since then, and the interest changed from consideration of the crowd evacuation as a whole to consideration of the microscopic motion of the individuals in the crowd evacuation by establishing individual movement rules. Many micromodels have been developed to accurately predict crowd motion and interpret a variety of self-organized phenomena in crowd motion and simulate hypothetical evacuation scenarios.

In this book, a microscopic particle-based model which is the distinct element method (DEM)-based multiagent model was utilized to formulate the crowd motion and simulate the uncertain behavior of evacuees and visualize in computer graphic environments as an application for crowd evacuation. The aim of the simulation is to reproduce realistic scenario of an evacuation process evolving in real time in dynamics virtual 3D environment. This book is written for beginners in this area of research and researchers who are advanced in some aspects of crowd dynamics and want to acquire DEM approach in simulating crowd dynamics.

Chapter Organization

This book is composed of five chapters. Chapter 1 provides an introduction to the crowd dynamics modeling and DEM-based crowd behavior simulator for evacuation simulation. Chapter 2 implies two subjects which are the fundamental of the distinct element method (DEM) and the framework of the DEM-based Crowd Behavior Simulator for Disaster Evacuation (CBS-DE). The fundamentals of the DEM presented in Sect. 2.2 are based on circular (two-dimension) or spherical (three-dimension) shaped rigid particles with deformable contact formulation. Meanwhile, the framework of the DEM-based CBS-DE is outlined in Sect. 2.3.

In this book, the development work was produced by extending the existing CBS-DE by introducing switching action behavior model. The mathematical modeling description of the switching action is presented in Chap. 3. Chapter 3 also reports on the data collections and methodology employed. The findings of the switching action behavior study are presented in Chap. 4. Finally, in Chap. 5, the conclusions are provided, and future works are discussed.

Nibong Tebal, Malaysia   Noorhazlinda Abd Rahman

Acknowledgements

My special thanks to *Prof. Dr. Hitoshi Gotoh* from Kyoto University for giving me precious opportunities working at his laboratory during my doctoral study. I am pleased to recognize *Assoc. Prof. Dr. Eiji Harada* from Kyoto University, for his valuable advice, constructive criticism, and his extensive discussion around my work. I also would like to acknowledge *School of Civil Engineering, Universiti Sains Malaysia* for full support in terms of moral and financial throughout my Ph.D. study and career. And, finally, I also wish to recognize my husband, *Zainal* and my children: *Idora, Amirul, Aleya,* and *Adib*.

Contents

Chapter 1
Introduction

1.1 Modeling and Simulation of Human Crowd Dynamics

For the past four decades, the development of different kinds of models can be seen to simulate complex crowd behavior. Modeling and simulation have become the best choice of methodology in understanding and predicting phenomena in a crowd by the replications. The replication of a phenomenon via mathematical model can demonstrates the dynamics of the phenomenon via simulation. This medium subsequently provides significant advantages that permit investigating the phenomenon behavior in an artificial way, which is often either not possible or too risky in the real world. The crucial traits in human crowd behavior and the emergence of several empirically observed collective behavior in a crowd could be reproduced and explained. By means of computer simulations, application in designing and planning pedestrian facilities and crowd evacuation could be performed.

The chronology of the development of crowd models can be traced back in 1970s wherein a macro modeling approach, known as the fluid dynamics-crowd flow model was first proposed by Henderson [1]. Henderson [1] conjectured the flow of people along a channel behaves similar to gasses or fluids. In his model the characteristics of crowd motion, such as crowd density and local speed, were described homogeneously. Henderson's phenomenological fluid dynamics approach had been improved by Helbing [2] by describing a fluid dynamics of the collective movement of pedestrians on a basis of a Boltzmann-like-gas-kinetic model. Unlike to Henderson's approach, Helbing [2] developed a specific theory for pedestrians without making use of unrealistic conservation assumptions. However, in fluid dynamics models the reality of pedestrian movement by allowing pedestrians to move around in an unrestricted manner was not reflected.

In 1990s, micro methods had gotten attention in modeling the crowd behavior. The micro methods concern with the motion characteristics of the individuals in the crowd. Each individual is tracked as it interacts with other individuals and the

N. Abd Rahman, *Crowd Behavior Simulation of Pedestrians During Evacuation Process*, SpringerBriefs in Computational Intelligence, https://doi.org/10.1007/978-981-13-1846-7_1

environment. Interactions are governed by contact between individuals and environments. Many micro models have been developed to accurately predict crowd motion and interpret a variety of self-organized phenomena in crowd motion. Typical micro models include cellular automata models [e.g., 3–8], social force models [e.g., 9–14], agent-based models [e.g., 15–19], and other particle-based models like DEM-based model [e.g., 20–24]. Currently, the DEM-based model in studying crowd dynamic still not many and more can be explored.

The models based on the aforementioned approaches (macro and micro approaches) are computational models, which can give numerical simulations of crowd behavior. Most models are application based, focusing on different aspects of crowd behavior. The use of modeling and simulation can provide an insight into the nature of the crowd behavior in which the solution of certain matters could be led. With the simulation, it could prove or disprove a variety of relationships observed in a crowd. A model is becoming a valuable research tool and planning tool these days. As such, a reliable model to produce a realistic simulation of human crowd behavior has become an important and challenging task.

1.2 DEM-Based Multi-agent Model

In this contribution, a microscopic particle-based model which is the Distinct Element Method (DEM)-based multi-agent model (hereinafter called DEM-based model) is employed. The DEM-based model is capable in modeling each individual distinctly by tracking the trajectory and rotation of each individual in the domain. Subsequently, the individual's position and orientation is evaluated through the calculation of interactions between individuals and between individual and boundary.

For many years, the DEM has been used extensively in modeling the granular flow, which has an analogy with a person flow. A simulator called Crowd Behavior Simulator for Disaster Evacuation (CBS-DE) has been developed by employing the DEM. The CBS-DE is a multi-agent crowd behavior simulator of microscopic individual-based model. It is designed with two important parts. The first part is a mathematical model, in which the behavior of the crowd is computed iteratively with advancing time. The second part is a visualization part, where the computed behavior in the first part is displayed. The model treats an individual behavior in a crowd directly, then simulates the crowd motion, and visualize in computer graphic environments as an application for crowd evacuation. The aim of the simulation is to reproduce realistic scenario of an evacuation process evolving in real-time in dynamics virtual 3D environment. The Autodesk® MAYA® 2010 (MAYA) environment for computer graphic movie is utilized in the CBS-DE.

In this contribution, the CBS-DE, which has been used in the previous studies [24–26] is employed and extended accordingly in order to model uncertain behavior of evacuees in a crowd. The uncertain behavior concern in this contribution is switching action behavior. New mathematical model of switching action behavior is formulated and introduced into the CBS-DE. Further, in simulating evacuations process, the newly developed model is demonstrated in different conjectural scenarios of evacuation process. In relation to the crowd evacuation, an observation of the evacuation process is conducted at the Langkawi International Airport (the LIA), Malaysia, as a pilot project.

1.3 Switching Action Behavior

The understanding of switching action behavior is necessary in order to develop a mathematical model. To the best of knowledge of this behavior, the switching action behavior has never been considered in any current crowd behavior models. This trait is believed to have significant effects in the crowd behavior itself and also in the evacuation process.

The switching action behavior in a crowd is proposed in this contribution in order to study the phenomenon of changing in a destination that may occur during the evacuation process. The mechanism of switching action behavior is proposed by formulating the switching action function through convolution integral of two functions. Further, the proposed model is employed to witness the performance of the switching action behavior model on different conjectural scenarios in the evacuation process with a change in destination. The details formulation and demonstration of the switching action behavior are discussed in the next chapters of the book.

References

1. Henderson LF (1974) On the fluid mechanics of human crowd motion. Transp Res 8(6):509–515
2. Helbing D (1992) A fluid dynamics model for a movement of pedestrians. Complex Syst 6:391–415
3. Zhao HT, Yang S, Chen XX (2016) Cellular automata model for urban road traffic flow considering pedestrian crossing street. Physica A 462:1301–1313. https://doi.org/10.1016/j.phys a.2016.06.146
4. Tao YZ, Dong LY (2016) Investigation on lane-formation in pedestrian flow with a new cellular automaton model. J Hydrodyn Ser B 28(5):794–800. https://doi.org/10.1016/S1001-6058(16) 60681-9
5. Kaji M, Inohara T (2017) Cellular automaton simulation of unidirectional pedestrians flow in a corridor to reproduce the unique velocity profile of Hagen-Poiseuille flow. Physica A 467:85–95
6. Li X, Guo F, Kuang H, Zhou H (2017) Effect of psychological tension on pedestrian counter flow via an extended cost potential field cellular automaton model. Physica A 487:47–57. https://doi.org/10.1016/j.physa.2017.05.070
7. Jin CJ, Jiang R, Yin JL, Dong LY, Li D (2017) Simulating bi-directional pedestrian flow in a cellular automaton model considering the body-turning behavior. Physica A 482:666–681. https://doi.org/10.1016/j.physa.2017.04.117
8. Hu J, You L, Zhang H, Wei J, Guo Y (2018) Study on queuing behavior in pedestrian evacuation by extended cellular automata model. Physica A 489:112–127. https://doi.org/10.1016/j.phys a.2017.07.004

9. Zeng W, Chen P, Nakamura H, Iryo-Asano M (2014) Application of social force model to pedestrian behavior analysis at signalized crosswalk. Transp Res Part C Emerg Technol 40:143–159. https://doi.org/10.1016/j.trc.2014.01.007

10. Johansson F, Peterson A, Tapani A (2015) Waiting pedestrians in the social force model. Physica A 419:95–107. https://doi.org/10.1016/j.physa.2014.10.003

11. Yang X, Dong H, Wang Q, Chen Y, Hu X (2014) Guided crowd dynamics via modified social force model. Physica A 411:63–73. https://doi.org/10.1016/j.physa.2014.05.068

12. Hou L, Liu JG, Pan X, Wang BH (2014) A social force evacuation model with the leadership effect. Physica A 400:93–99. https://doi.org/10.1016/j.physa.2013.12.049

13. Han Y, Liu H (2017) Modified social force model based on information transmission toward crowd evacuation simulation. Physica A 469:499–509. https://doi.org/10.1016/j.physa.2016.11.014

14. Zhang H, Liu H, Qin X, Liu B (2018) Modified two-layer social force model for emergency earthquake evacuation. Physica A 492:1107–1119. https://doi.org/10.1016/j.physa.2017.11.041

15. Collins A, Elzie T, Frydenlund E, Robinson RM (2014) Do groups matter? An agent-based modeling approach to pedestrian egress. Transp Res Procedia 2:430–435. https://doi.org/10.1016/j.trpro.2014.09.051

16. Bernardini G, D'Orazio M, Quagliarini E, Spalazzi L (2014) An agent-based model for earthquake pedestrians' evacuation simulation in urban scenarios. Transp Res Procedia 2:255–263. https://doi.org/10.1016/j.trpro.2014.09.050

17. Shaaban K, Abdel-Warith K (2017) Agent-based modeling of pedestrian behavior at an unmarked midblock crossing. Procedia Comput Sci 109:26–33. https://doi.org/10.1016/j.procs.2017.05.291

18. Fujii H, Uchida H, Yoshimura S (2017) Agent-based simulation framework for mixed traffic of cars, pedestrians and trams. Transp Res Part C Emerg Technol 85:234–248. https://doi.org/10.1016/j.trc.2017.09.018

19. Karbovskii V, Voloshin D, Karsakov A, Bezgodov A, Gershenson C (2018) Multimodel agent-based simulation environment for mass-gatherings and pedestrian dynamics. Future Gener Comput Syst 79(Part 1):155–165. https://doi.org/10.1016/j.future.2016.10.002

20. Langston P, Masling R, Asmar B (2006) Crowd dynamics discrete element multi-circle model. Saf Sci 44(5):395–417. https://doi.org/10.1016/j.ssci.2005.11.007

21. Smith A, James C, Jones R, Langston P, Lester E, Drury J (2009) Modelling contra-flow in crowd dynamics DEM simulation. Saf Sci 47(3):395–404. https://doi.org/10.1016/j.ssci.2008.05.006

22. Singh H, Arter R, Dodd L, Langston P, Lester E, Drury J (2009) Modelling subgroup behaviour in crowd dynamics DEM simulation. Appl Math Model 33(12):4408–4423. https://doi.org/10.1016/j.apm.2009.03.020

23. Kiyono J, Mori N (2004) Simulation of emergency evacuation behavior during disaster by use of elliptic distinct element method. In: 13th World conference on earthquake engineering, 13 WCEE, Vancouver, B.C., Canada, Paper No. 134

24. Gotoh H, Harada E, Andoh E (2012) Simulation of pedestrian contra-flow by multi-agent DEM model with self-evasive action model. Saf Sci 50(2):326–332. https://doi.org/10.1016/j.ssci.2011.09.009

25. Gotoh H, Harada E, Muruyama Y, Takahishi K, Ohniwa K (2008) Contribution of crowd refuge simulator to town planning against tsunami flood. In: Proceeding of coastal engineering, JSCE, vol 55, pp 1371–1375 (in Japanese). https://doi.org/10.2208/proce1989.55.1371

26. Gotoh H, Harada E, Ohniwa K (2009) Numerical simulation of coastal town planning against tsunami by DEM-based human behavior simulator. In: Proceedings of the nineteenth international offshore and polar engineering conference, ISOPE-2009, Osaka, Japan, pp 1248–1252

Chapter 2
DEM-Based Crowd Behavior Simulator

2.1 Background

The Distinct Element Method (DEM) is a set of computational modeling technique that treat a medium as a discontinuum in contrast to other established methods, like FEM, that treats a medium as a continuum. It reveals discontinuous mechanical behavior in engineering and applied science problems. The supremacy of this method is in the dynamics model, including the motion and mechanical interactions between particles in any simulation and provides a detailed description of the forces acting on each particle. In the DEM, the medium is modeled as an assembly of rigid distinct elements and these elements are connected to each other by virtual spring and dashpot. It is assumed that the behavior of each element is governed by the body force and contact force among elements.

For many years, the DEM has been used extensively in modeling the granular flow, which has an analogy with a crowd flow. Thus, DEM-based crowd behavior simulator, named Crowd Behavior Simulator for Disaster Evacuation (CBS-DE) has been developed. In this simulator, a person is represented as a single cylindrical element and consequently a crowd is regarded as an assembly of cylindrical elements. By using the information acquired from the vision of a person, namely the distance between adjacent persons and person to wall boundary, human behavior is modeled. A set of computational methods, including autonomous and interaction algorithms are incorporated into CBS-DE to simulate different traits of human behavior like the collision avoidance, slow-down effect, overtaking and self-organize pattern like lane formation. The computational framework of CBS-DE allows easy extension by inclusion of new additional traits of crowd behavior later on.

Previously, CBS-DE was developed to study tsunami flood refuge [1] and had been employed in the study of coastal town planning [2] and evacuation process [3]. Later, CBS-DE was improved by introducing a dynamic model of self-evasive action [4]. CBS-DE with self-evasive action model can treat collision avoidance

N. Abd Rahman, *Crowd Behavior Simulation of Pedestrians During Evacuation Process*, SpringerBriefs in Computational Intelligence, https://doi.org/10.1007/978-981-13-1846-7_2

from other adjacent pedestrians and an alignment following preceding pedestrians. The simulation of a highly crowded pedestrian contra-flow at a crossing with the self-evasive action model showed satisfactory agreement with the observation.

In reference to the objective of this contribution, CBS-DE with self-evasive action model, which has been shown its reasonable performance in simulating crowd behavior in the previous study, is adopted in modeling and simulating the switching action behavior in a crowd.

2.2 Fundamental of Distinct Element Method (DEM)

The descriptions of the DEM theory in this book are based on papers authored by Gotoh et al. [5], Kawamura and Kobayashi [6] and chapters in books by Harada and Gotoh [7], O'Sullivan [8].

2.2.1 An Overview

The DEM is known as a suitable method for the simulation of the dynamic behavior of an assembly of particles. This approach explicitly provides the mechanical behavior of the individual particles and their contacts. Its computational modeling framework allows finite displacements and rotations of discrete particles and recognizes new contacts automatically.

Conceptually, particles move according to Newton's second law of motion. The fundamental particles are rigid and possess mass and moment of inertia which can move independently of each other and can translate and rotate. When particles contact with one another, 'new' contact established and 'old' contact is released. This cause changes in the contact status and interaction forces influence the subsequent movement of bodies. This concept forms the basis of DEM approach.

The calculation sequence of DEM simulation is illustrated in Fig. 2.1. To perform a DEM simulation, initially, a system geometry, including particle positions and boundary conditions are initialized. Contact model parameters like stiffness, viscosity, and friction coefficient are also specified. In each computational time-step, the particles in contact are detected. The magnitudes of inter-particle forces related to the distance between particles in contact are calculated. Subsequently, the resultant force acting on each particle can be determined. Then, the translational and rotational accelerations of particles can be calculated. The displacement and rotation of particles over the current time-step is given by time integration. By using these incremental displacements and rotations, the position and orientation of particles are updated. In the next time-step, the contact forces are calculated by using this updated geometry and the series of calculations are repeated.

Fig. 2.1 Schematic diagram
of the flowchart in the DEM
simulation

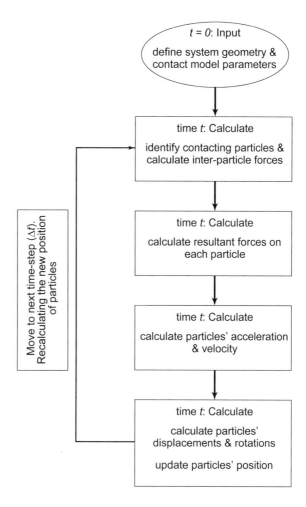

2.2.2   Computing Method

The DEM models particles distinctly as a group of rigid bodies and govern the behavior of each particle by translational and rotational equations of motion as follows:

$$F = ma \tag{2.1}$$

$$T = I\alpha \tag{2.2}$$

where F reflects internal and external forces, T is a torque, m is a mass of a particle, I is a moment of inertia of a particle, a refers an acceleration of a particle and α is an angular acceleration of a particle.

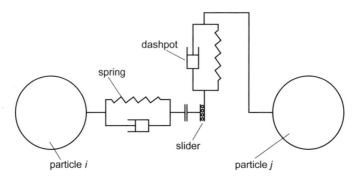

Fig. 2.2 Spring-dashpot system

The interaction between particles in contact is evaluated by the spring-dashpot system as depicted in Fig. 2.2. The spring-dashpot system reflects the mechanical system consists of a spring, a dashpot and a friction slider which respectively arranged in the normal and tangential directions to the tangential plane having a common contact point on particles i and j as shown in Fig. 2.2.

In consideration of interaction forces between particles, Eqs. 2.1 and 2.2 are expanded as follows:

$$m\ddot{x} + c\dot{x} + kx = 0 \tag{2.3}$$

$$I\dot{\omega} + cr^2\dot{\omega} + kr^2\omega = 0 \tag{2.4}$$

in which, x is a translational displacement, ω is a rotational displacement of a particle, "·" indicates a time-derivative, c is a dashpot damping coefficient, k is a spring elastic constant and r is a radius of a particle. Equations 2.3 and 2.4 above are integrated explicitly in the time domain. The acceleration of each particle is computed explicitly by considering the information for the velocity and position in the previous time-step as follows:

$$m\ddot{x} = -c\dot{x} - kx \tag{2.5}$$

To solve Eq. 2.5, extraction of particles in contact is needed to assess the inter-particle force, and this is achieved by considering the distance between two particles in contact as shown in Fig. 2.3.

As shown in Fig. 2.3, the distance between two particles is computed as follows:

$$r_{ij} = \sqrt{\left(x_i - x_j\right)^2 + \left(y_i - y_j\right)^2} \tag{2.6}$$

in which r_{ij} is the distance between particle i and particle j, and x, and y denote the coordinates of the particle. The two particles are in contact with one another if the sum of the two radii of particles i and j exceeds the distance r_{ij}:

Fig. 2.3 Contact of two particles

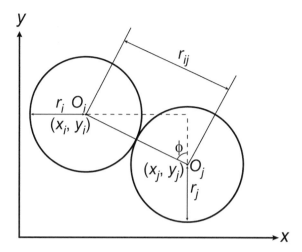

$$r_i + r_j \geq r_{ij} \tag{2.7}$$

in which r_i and r_j are radii for particle i and j, respectively. The relation between particles in contact; particles i and j, and its angle ϕ shown in Fig. 2.3 can be calculated in the following equations:

$$\sin \phi = -(x_i - x_j)/r_{ij} \tag{2.8}$$
$$\cos \phi = -(y_i - y_j)/r_{ij} \tag{2.9}$$

The relative displacement in the unit computational time-step between two particles can be described as follows:

$$\Delta r^n = (\Delta x_i - \Delta x_j)\cos \phi + (\Delta y_i - \Delta y_j)\sin \phi \tag{2.10}$$
$$\Delta r^t = -(\Delta x_i - \Delta x_j)\sin \phi + (\Delta y_i - \Delta y_j)\cos \phi + (r_i \Delta \omega_i + r_j \Delta \omega_j) \tag{2.11}$$

where Δr is the component of a relative displacement between particles i and j, superscripts n and t denote normal and tangential directions, respectively, Δx and Δy are displacements of particles in x and y directions, respectively, $\Delta \omega$ is a rotational angle of a particle and r is a radius of a particle.

Contact forces between two particles are described by using relative displacement and velocity in normal and tangential directions since the two particles are connected by a spring and a dashpot in two directions as illustrated in Fig. 2.2. The normal contact force is described as follows:

$$\Delta e^n = k^n \Delta r^n \tag{2.12}$$
$$e^n = e^n_{pre} + \Delta e^n \tag{2.13}$$

$$\Delta d^n = c^n \frac{\Delta r^n}{\Delta t} \tag{2.14}$$

$$d^n = \Delta d^n \tag{2.15}$$

$$f^n = e^n + d^n \tag{2.16}$$

where Δe^n is the increment of the normal force due to a spring in the current time-step, e^n is the normal force due to a spring, k^n is a spring constant in the normal direction, Δd^n is the increment of the normal force due to a dashpot in the current time-step, d^n is the normal force due to a dashpot, c^n is a damping coefficient in the normal direction and f^n is the normal interaction force in the current time-step. Meanwhile, Δt denotes the time marching step, and the subscript '*pre*' indicates the previous time-step.

The tangential contact force is deduced in a similar way to the normal contact force:

$$\Delta e^t = k^t \Delta r^t \tag{2.17}$$

$$e^t = e^t_{pre} + \Delta e^t \tag{2.18}$$

$$\Delta d^t = c^t \frac{\Delta r^t}{\Delta t} \tag{2.19}$$

$$d^t = \Delta d^t \tag{2.20}$$

$$f^t = e^t + d^t \tag{2.21}$$

where Δe^t is the increment of the tangential force due to a spring in the current time-step, e^t is the tangential force due to a spring, k^t is a spring constant in the tangential direction, Δd^t is the increment of the tangential force due to a dashpot in the current time-step, d^t is the tangential force due to a dashpot, c^t is a viscosity coefficient in the tangential direction and f^t is the tangential interaction force.

The total forces acting on each particle are obtained by summing all forces in each axis direction:

$$F_i^x = \sum_j \left\{ -f^n \cos\phi + f^t \sin\phi \right\} \tag{2.22}$$

$$F_i^y = \sum_j \left\{ -f^n \sin\phi - f^t \cos\phi \right\} \tag{2.23}$$

$$M_i = -r_i \sum_j f^t \tag{2.24}$$

where F_i^x is the total force component in the x direction, F_i^y is the total force component in the y direction, and M_i is the total moment. The acceleration of a particle i is obtained by dividing those forces by the mass or moment of inertia.

$$a_x = \frac{F_i^x}{m} \tag{2.25}$$

$$a_y = \frac{F_i^y}{m} \tag{2.26}$$

$$\ddot{\omega} = \frac{M_i}{I} \tag{2.27}$$

Euler explicit time integration is applied to Eqs. 2.25, 2.26 and 2.27. And, the velocity and position of a particle i are computed.

2.2.3 Contact Model Parameters

For procedures described in the above section, the spring constant and coefficient of damping are necessary. Hertzian contact theory is employed to determine model parameters. In Hertzian contact theory, the overlap δ between two spheres is related to the contact force P in the normal direction, Young's modulus E and Poisson's ratio ν. Figure 2.4 illustrates the relation between two particles in contact.

The non-linear relation between the contact force and overlap is given as the following:

Fig. 2.4 Contact between two particles

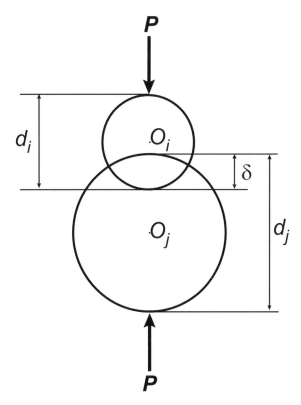

$$P = k^n \delta^{3/2} \tag{2.28}$$

And, k^n is the spring constant in the normal direction which is given by:

$$k^n = \frac{\sqrt{2}}{3} \sqrt{\frac{d_i d_j}{d_i + d_j}} \frac{E}{1 - v^2} \tag{2.29}$$

in which d is the diameter of a particle. The spring constant in the tangential direction, k^t is estimated by using the ratio of Young's modulus E to the shear modulus G.

$$s_0 = \frac{k^t}{k^n} = \frac{G}{E} = \frac{1}{2(1 + v)} \tag{2.30}$$

where s_0 is the damping rate.

Further, the coefficient of damping is defined to achieve critical damping condition of the spring-dashpot system for the one degree of freedom. From Eq. 2.3, we have:

$$\ddot{x} + \frac{c}{m}\dot{x} + \frac{k}{m}x = 0 \tag{2.31}$$

Let

$$\omega^2 = \frac{k}{m}; 2\zeta = \frac{c}{m} \tag{2.32}$$

Using these notations, we can write:

$$\lambda^2 + 2\zeta\lambda + \omega^2 = 0 \tag{2.33}$$

Considering the double root condition of Eq. 2.33, the critical damping coefficient is obtained. The damping coefficient in the normal direction is given by Eq. 2.34:

$$c^n = 2\sqrt{mk^n} \tag{2.34}$$

and the damping coefficient in the tangential direction is given as:

$$c^t = c^n\sqrt{s_0} \tag{2.35}$$

By considering the numerical stability, the time marching step Δt should be set appropriately and sufficiently small. The time marching step is deduced from the relation of the natural period T in the spring-mass system as follows:

$$\Delta t = \frac{T}{\varphi}; T = 2\pi\sqrt{\frac{m}{k^n}} \tag{2.36}$$

in which φ is a parameter in the range $10 \le \phi \le 20$.

2.3 Mathematical Modeling of CBS-DE

In this sub-chapter, the framework of mathematical modeling of the DEM-based CBS-DE is elucidated in detail, in which reference is made based on several papers authored by Gotoh et al. [1–3]. Further, the self-evasive action model in the existing CBS-DE is described in reference to the formulation presented in Gotoh et al. [4].

2.3.1 Governing Equations

The motion of a person element in contact with neighboring elements is described in accordance with Newton's law of motion. Each person element is governed by translational and rotational equations of motion. The combination of an autonomous driving force, repulsive forces and the self-evasive force describe the motion of a person. The autonomous driving force reflects the motivation of a person to move in a prescribed walking velocity and orientation while repulsive forces reflect inter-element (contact) forces due to collisions, and the self-evasive force treats collision avoidance and person alignment. Hence, the motion of the person i in CBS-DE is written as follows:

$$m_{hi}\dot{v}_{hi} = F_{inhi} + F_{awhi} + F_{sehi} \tag{2.37}$$

$$I_{hi}\dot{\omega}_{hi} = T_{hi} \tag{2.38}$$

where m_{hi} and I_{hi} are two parameters that refer to the mass and moment of inertia of the person i, respectively. Both parameters are given based on the representation of a person as a cylindrical element as shown respectively in Eqs. 2.39 and 2.40. In the above equations, v_{hi} is the velocity of the person i, ω_{hi} is the angular velocity of the person i, "$\cdot$" indicates a time-derivative, F_{inhi} is the inter-element (contact) force acting on the person i, F_{awhi} is the autonomous walking force of the person i, F_{sehi} is the self-evasive force acting on the person i and T_{hi} is the torque acting on the person i. The behavior of the person element is computed explicitly by numerical integration of Eqs. 2.37 and 2.38.

Each person is simulated as a cylinder, thus m_{hi} and I_{hi} can be written as in Eqs. 2.39 and 2.40, respectively:

$$m_{hi} = \frac{1}{4}\varepsilon_{hi}\sigma_{hi}B_{hi}\pi d_{hi}^2 \tag{2.39}$$

$$I_{hi} = \frac{1}{32}\varepsilon_{hi}\sigma_{hi}B_{hi}\pi d_{hi}^4 \tag{2.40}$$

where ε_{hi} is the volume coefficient concerning the volume difference between the cylindrical element and the actual person, σ_{hi} is the density of the person i, B_{hi} is the

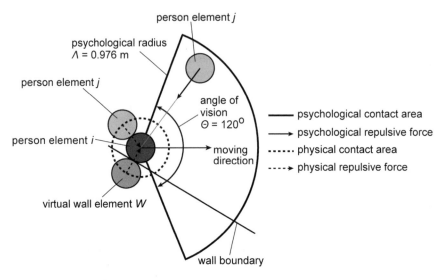

Fig. 2.5 Schematic diagram of the perception domain of a person i

body height of the person i and d_{hi} is the diameter of the person i. In this model, by considering $m_{hi} = 56.5$ kg, $\sigma_{hi} = 980$ kg/m^3, $B_{hi} = 164.7$ cm and $d_{hi} = 0.379$ m, the ε_{hi} is set as 0.31.

The repulsive force describes "person-to-person" and "person-to-wall boundary" interactions. It consists of physical and psychological ones. The estimation of the repulsive force is made by the Voigt model (the spring-dashpot system) in reference to the perception domain (the personal space) as shown in Fig. 2.5. Meanwhile, the self-evasive force represents the effect of psychological repulsive force. It is determined by considering the predicted position of surrounding persons after Δt_f seconds.

2.3.2   Representation of the Perception Domain

Vision is the primary source of information to control persons' motion. In CBS-DE, the vision of a person i is represented by a perception domain which reflects as a personal space of a person i. Figure 2.5 is a model of the vision (vision fan) showing the perception domain of a person i, which is symmetric to the moving direction of each person. Psychologically, people tend to keep a constant distance with each other when they walk. Hence, in CBS-DE, a constant distance is adopted for each person i which is called a psychological radius Λ with 0.976 m length. This is an empirical value that was observed by Kiyono et al. [9] as the average of the relative distance between persons stopping at the red light.

The perception domain is also used to evaluate the interactions of person i to his/her adjacent persons and to his/her wall boundary. CBS-DE assumes that persons are not affected by anyone behind them and pay less attention to what occur behind them. Hence, the person i only pays attention to persons and wall boundary at the front. For this reason, the vision field of person i, named angle of vision Θ, is decided and it ranges in the degree from $-60°$ to $60°$ with respect to the moving direction of the person i.

2.3.3 Inter-element Force

If the relative distance between the centroids of the person i and j satisfies the following conditions (Eqs. 2.41 and 2.42), physical and psychological repulsive forces between persons i and j are activated,

$$|r_i - r_j| \leq \frac{d_{hi} + d_{hj}}{2} \tag{2.41}$$

$$|r_i - r_j| \leq \Lambda \tag{2.42}$$

Meanwhile, the physical repulsive force between person i and virtual wall element W acts under the following condition:

$$|r_i - r_W| \leq \frac{d_{hi} + d_W}{2} \tag{2.43}$$

The virtual wall element W is positioned on the wall so as to have tangential contact with the person i. In the above equations, r_i is the positional vector of the person i, r_W is the positional vector of the virtual wall element W and d_W is the diameter of the virtual wall element W with the same diameter as a person element (0.379 m).

The inter-element force is determined by the spring-dashpot system. The total interacting force acting on the person i is described as follows:

$$F_{inhi} = F_{ij} + F_{iW} + F_{psij} \tag{2.44}$$

$$F_{ij} = \sum_{j(\neq i)} f_{ij}, \quad F_{iW} = \sum_{W} f_{iW}, \quad F_{psij} = \sum_{j(\neq i)} f_{psij} \tag{2.45}$$

$$T_{hi} = \frac{1}{2} \left\{ \sum_{j(\neq i)} (r_j - r_i) \times f_{ij} + \sum_{W} (r_W - r_i) \times f_{iW} \right\} \tag{2.46}$$

where F_{ij} is the net physical repulsive force between persons i and j, F_{iW} is the net physical repulsive force between person i and virtual wall element W, F_{psij} is the net psychological repulsive force between persons i and j, f_{ij} is the local physical repulsive force between persons i and j, f_{iW} is the local physical repulsive force between person i and virtual wall element W, and f_{psij} is the local psychological repulsive force between persons i and j.

The local repulsive forces, f_{ij}, f_{iW} and f_{psij} are quantified as follows:

$$\boldsymbol{f}_* = f_*^n \boldsymbol{n} + f_*^t \boldsymbol{t} \tag{2.47}$$

$$f_*^n = (e + d)_*^n, f_*^t = (e + d)_*^t \tag{2.48}$$

$$e_*^n = \left(e_{pre} + k\delta\right)_*^n; e_*^t = \left(e_{pre} + k\delta\right)_*^t \tag{2.49}$$

$$d_*^n = \left(c\frac{\delta}{\Delta t}\right)_*^n; d_*^t = \left(c\frac{\delta}{\Delta t}\right)_*^t \tag{2.50}$$

where $\boldsymbol{n}$ and $\boldsymbol{t}$ are the unit vector in the normal and tangential directions, respectively, the superscripts n and t are the indicators in the normal and tangential directions, respectively, f is the inter-element force, e is the component of the inter-element force due to the spring, d is the component of the inter-element force due to the dashpot, δ is the relative displacement between contact elements, Δt is the time marching step, e_{pre} is the component of the inter-element force due to the spring at the previous time step, k and c are the model parameters of the spring and dashpot, respectively, and the subscript "*" indicates any of the following subscripts: ij, iW and $psij$.

Each person is assumed to walk individually. Hence, a joint without resistance against the tensile force is introduced in the normal direction. Moreover, a slider joint is allocated to represent frictional effects. Thus, according to the contact condition between persons, the inter-element forces f_* given in Eq. (2.47) is modified as:

$$e_*^n < 0 \text{ then } \boldsymbol{f}_* = \boldsymbol{0} \tag{2.51}$$

$$\left|e_*^t\right| > \mu e_*^n \text{ then } f^t = \mu \cdot \text{Sign}\left[e_*^n, e_*^t\right] \tag{2.52}$$

where μ $(=0.577)$ is the friction coefficient. Sign [A, B] represents the absolute value of A with the sign of B.

2.3.4 Autonomous Driving Force

Under free boundary condition or without any neighboring effects, each person i with the mass m_{hi} is driven to a given direction with a certain specific equilibrium velocity. The equilibrium velocity is subjected to the congestion condition in a person's vision (the perception domain). Hence, the person i correspondingly tends to accelerate up to the specific equilibrium velocity by the autonomous driving force, which is given by:

$$F_{awhi} = m_{hi}a \qquad (2.53)$$

where a is the acceleration vector. A correlation between the number density of persons in the perception domain and the specific equilibrium velocity is given by:

$$v_{max} = v_{limit} - \gamma \cdot c_k \qquad (2.54)$$

where v_{limit} is the specific equilibrium velocity in a free walking condition, γ (=0.54) is the parameter for an attenuation effect due to the congestion condition in a person's vision (the perception domain) and c_k is the number density of persons inside the perception domain.

2.3.5 Self-evasive Action Force

The self-evasive force is determined by considering the predicted position of surrounding persons after Δt_f seconds as shown in Fig. 2.6. Gotoh et al. [4] described the self-evasive action force as follows:

$$F_{sehi} = \begin{cases} m_{hi} \sum\limits_{j(\neq i)}^{N_k} \kappa \dfrac{(v_{hi} - v_{hj}) \cdot e_{vi}}{\Delta t_f} \left(\hat{e}_{ij} \times e_{vi}\right) \times e_{vi} & \text{when } \hat{e}_{ij} \times e_{vi} \neq 0 \\[3mm] m_{hi} \sum\limits_{j(\neq i)}^{N_k} \kappa \dfrac{(v_{hi} - v_{hj}) \cdot e_{vi}}{\Delta t_f} \left(\hat{e}_x \times e_y\right) \times e_{vi} & \text{when } \hat{e}_{ij} \times e_{vi} = 0 \end{cases} \qquad (2.55)$$

$$\kappa = \alpha_{se} \frac{\cos \hat{\phi}_{ij}}{|\hat{r}_{ij}|/\lambda_a} \; ; \; \cos \hat{\phi}_{ij} = e_{vi} \cdot \hat{e}_{ij} \qquad (2.56)$$

$$\hat{r}_{ij} = \hat{r}_j - r_i \; ; \hat{r}_j = r_j + v_{hj}\Delta t_f \qquad (2.57)$$

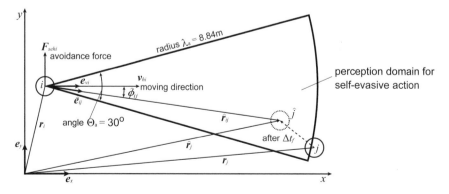

Fig. 2.6 Perception domain of the self-evasive action

where N_k is the total number of persons in the perception domain of the self-evasive action shown in Fig. 2.6, Δt_f denotes the time for the self-evasive action, e_{vi} is the unit vector in the v_{hi} direction, $\hat{e}_{ij}$ is the unit vector in the $\hat{r}_{ij}$ direction, e_x and e_y are the unit vectors in the x- and y-axis, respectively, α_{se} ($=4.5 \times 10^{-4}$) is the coefficient of the self-evasive action, $\hat{r}_{ij}$ is the predicted positional vector of the person j after Δt_f, and λ_a is the radius of the perception domain for the self-evasive action. In the previous simulation, Δt_f is given as 2.0 s. Equation (2.55) indicates that the avoidance force acts on the person i in case of increasing relative travelling velocity of person j to persons i; meanwhile, the alignment force acts on the person i in case of decreasing relative velocity of persons j to persons i. As for the singularity treatment, when the unit vector $\hat{e}_{ij}$ equal to the unit vector e_{vi}, the avoidance force F_{sehi} acts on the person i in the right direction perpendicular to the travelling direction of the person i.

As mentioned earlier, CBS-DE with self-evasive action model has been demonstrated in the study of the pedestrian contra-flow in highly crowded situation at a crossing. Naturally, the contra-flow may occur during the evacuation process due to any unexpected situations. Consequently, the self-evasive action model is taken into account in this research.

2.3.6 Model Parameters

In performing the simulation by CBS-DE, the model parameters such as the acceleration for the driving force of a person, the walking speed of the evacuees, the spring and dashpot constants need to be calibrated. In this research, the model parameters in CBS-DE with self-evasive action model [4] are used. For the normal case, the acceleration is taken as 0.837 m/s^2. Meanwhile, the magnitude of specific equilibrium walking velocity, v_{limit} is taken as 1.47 m/s.

In the present study, the spring and dashpot constants using in the DEM are given according to the setup procedure proposed by Kiyono et al. [9, 10] as shown in Table 2.1. In their procedure, the normal spring constant is given first. Then, the tangential spring constant is calculated by multiplying 0.05 to the normal spring constant. The value of 0.05 was determined through trial and error by Kiyono et al. [9, 10]. Meanwhile, the damping coefficient is decided by considering the critical damping condition of the Voigt model for the one degree of freedom. Table 2.2 shows the setup procedure in determining the models constant introduced in Kiyono et al. [9, 10].

Table 2.1 Model parameters

Constant	Parameter	Value	Unit
Physical normal spring constant	k_{ij}^n	1.26×10^4	N/m
Physical tangential spring constant	k_{ij}^t	6.30×10^2	N/m
Physical normal dashpot constant	c_{ij}^n	1.69×10^3	N s/m
Physical tangential dashpot constant	c_{ij}^t	3.77×10^2	N s/m
Psychological normal spring constant	k_{psij}^n	6.01×10	N/m
Psychological tangential spring constant	k_{psij}^t	3.01	N/m
Psychological normal dashpot constant	c_{psij}^n	1.17×10^2	N s/m
Psychological tangential dashpot constant	c_{psij}^t	2.61×10	N s/m

Table 2.2 Setup procedure of models constant

		Physical repulsive force	Psychological repulsive force
Spring	k^n	k^n	$k^n = \dfrac{ma}{\Lambda - \frac{d_h}{2}}$
	k^t	$k^n \times 0.05$	$k^n \times 0.05$
Dashpot	c^n	$2\sqrt{mk^n}$	$2\sqrt{mk^n}$
	c^t	$2\sqrt{0.05mk^n}$	$2\sqrt{0.05mk^n}$

References

1. Gotoh H, Harada E, Kubo Y, Sakai T (2004) Particle-system model of the behavior of crowd in tsunami flood refuge. Ann J Coast Eng JSCE 51:1261–1265 (in Japanese)
2. Gotoh H, Harada E, Muruyama Y, Takahishi K, Ohniwa K (2008) Contribution of crowd refuge simulator to town planning against tsunami flood. In: Proceeding of Coastal Engineering, JSCE, vol 55, pp 1371–1375 (in Japanese). https://doi.org/10.2208/proce1989.55.1371
3. Gotoh H, Harada E, Ohniwa K (2009) Numerical simulation of coastal town planning against tsunami by DEM-based human behavior simulator. In: Proceedings of the nineteenth (2009) international offshore and polar engineering conference, ISOPE-2009, Osaka, Japan, pp 1248–1252
4. Gotoh H, Harada E, Andoh E (2012) Simulation of pedestrian contra-flow by multi-agent DEM model with self-evasive action model. Saf Sci 50(2):326–332. https://doi.org/10.1016/j.ssci.2011.09.009
5. Gotoh H, Sakai T (1997) Numerical simulation of sheetflow as granular material. J Waterw Port Coast Ocean Eng ASCE 123(6):329–336

6. Kawamura Y, Kobayashi Y (2012) Basic study on application of discrete element method for slope failure analysis. In: Proceeding of the 15th world conference on earthquake engineering 2012. Retrieved from http://www.iitk.ac.in/nicee/wcee/article/WCEE2012_5478.pdf
7. Harada E, Gotoh H (2013) Distinct element method. In: Gotoh H, Okayasu A, Watanabe Y (eds) Computational wave dynamics, advances series on ocean engineering. World Scientific Publiching Co. Pte. Ltd, Singapore, vol 37, pp 181–194
8. O'Sullivan C (2011) Particulate discrete element modeling, a geomechanics perspective. Appl Geotech 4
9. Kiyono J, Miura F, Takimoto K (1996) Simulation of emergency evacuation behavior during disaster by using distinct element method. J Struct Mech Earthquake Eng JSCE 537(I-35): 233–244 (in Japanese)
10. Kiyono J, Miura F, Yagi H (1998) Simulation of evacuation behavior in a disaster by distinct element method. J Struct Mech Earthquake Eng JSCE 591 (I-43):365–378 (in Japanese)

Chapter 3
Switching Action Behavior Model in Crowd Behavior Simulator

3.1 The Extension of Crowd Behavior Simulator

The Crowd Behavior Simulator for Disaster Evacuation (CBS-DE) based on the Distinct Element Method (DEM) is a microscopic individual-based model which can treat an individual behavior directly and suitable for the reproduction of details of an evacuation process. Not only the simulator has been designed for describing observed crowd behaviors, like collision avoidance between pedestrians or boundary, slow-down effect, overtaking, pedestrian alignment and lane formation, but also it has been used in practical application especially in disaster evacuation planning. As such, a set of human behavior model including autonomous and interaction algorithms are incorporated into CBS-DE whose computational framework will allow us to include new additional traits of crowd behavior with ease in the future.

In this contribution, a new additional trait of crowd behavior is studied in order to investigate their effects in evacuation simulation. That new additional trait of crowd behavior is switching action behavior. The new model is added in the existing CBS-DE. The extended CBS-DE is named as CBS-DE with switching action behavior model. The following subchapter will elucidate the mathematical formulation of the model.

N. Abd Rahman, *Crowd Behavior Simulation of Pedestrians During Evacuation Process*, SpringerBriefs in Computational Intelligence, https://doi.org/10.1007/978-981-13-1846-7_3

3.2 Switching Action Behavior Model

When the number of evacuees exceeds a space capacity of an evacuation area, people who cannot enter the evacuation area have to change their destination toward another evacuation area. The space capacity of the evacuation area is estimated based on the definition of the crowd density by Department for Culture, Media and Sports [1]. In order to simulate the change in destination, a switching action behavior model is developed. Up to now the phenomenon of people who has to change in destination has never been taken into account in any evacuation modeling. The significance of this factor in the simulation for evacuation is indisputable. For example, when the designated evacuation area is fully occupied with preceding evacuees; trailing evacuees have to change their directions to another evacuation area. Hence, switching action behavior may occur during evacuation situations. To represent the aforementioned phenomenon briefly, the switching action behavior was modeled, and the existing CBS-DE was extended by incorporating the switching action behavior algorithm. The mathematical formulation of the switching action behavior proposed in this study is based on the convolution integral of two functions.

3.2.1 Mathematical Modeling

In the current contribution, a switching action behavior model is developed for simulating the change in destination of a person i due to the limitation of space capacity of the evacuation area. The change in destination of a person i is dependent on the motion of other persons in the perception domain in the defined switching action time. The switching action time is given by the switching action function described by a convolution integral.

Figure 3.1 shows the flow chart of the newly developed switching action behavior model. Person i evacuates to the evacuation area until the space of the evacuation area is not available.

If the space is not available, in other words, the space capacity of the evacuation area is fully occupied, the person i need to change his/her destination to other evacuation area by changing the direction. To simulate this change in direction, the switching action behavior function is defined for each person i. To avoid an artifact of the flow of change in direction, which means a sudden change in direction, the change in direction of the person i is determined by reference to the motion of other persons in the perception domain in the defined switching action time. The switching action function is only activated at the time t when a person i satisfies the following conditions:

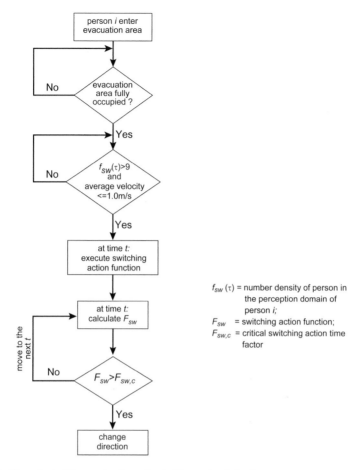

Fig. 3.1 Flow chart of the mechanism of switching action behavior

(i) the number density of persons in the perception domain of a person i is more than 9; and

(ii) the average walking velocity of these persons in the perception domain is less than or equal to 1.0 m/s.

The number of 9 persons is decided by considering the congestion density in the perception domain of a person i with a radius for switching action model 1.516 m ($=4d_{hi}$), as shown in Fig. 3.2.

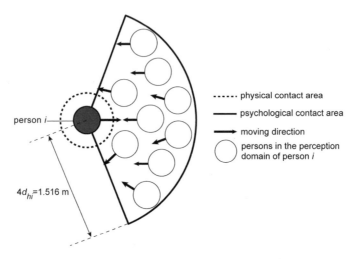

Fig. 3.2 Schematic illustration of the congestion in the perception domain of the person i

The switching action function for the person i is calculated by the following equations.

$$F_{sw}(t) = \int_{t-\alpha_{sw}}^{t} f_{sw}(\tau) g_{sw}(t - \tau) d\tau \qquad (3.1)$$

$$g(t) = e^{-t} \qquad (3.2)$$

where $F_{sw}(t)$ is the switching action function, f_{sw} is the input function, which indicates the number density of persons in the perception domain of the person i and g_{sw} is the unit response function. The response function reflects the delay of switching action. For the $g(t)$, a monotonous damping function with respect to time contributes a computational stability. In the present study, the exponentially-decaying function is employed as the $g(t)$.

The change in direction will only takes place once the switching action function exceeds the critical switching action time factor. The critical switching action time factor, $F_{sw,c}$ is introduced in this study, which varies according to the congestion and perception-response time α_{sw}. A perception-response time α_{sw} is a model parameter that has to be calibrated in order to obtain an appropriate switching action behavior. Once $F_{sw}(t)$ of the person i attains the critical switching action time factor $F_{sw,c}$, the person i will change his/her direction to the opposite direction. $F_{sw,c}$ was given by the assumption that 9 persons exist continuously in the perception domain in the perception-response time α_{sw} as follows:

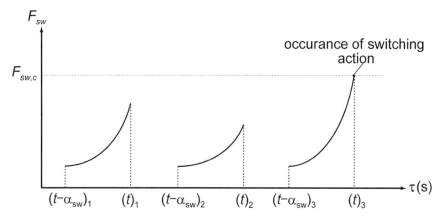

Fig. 3.3 Example of time series of the switching action function

$$F_{sw}(t) = 9 \int_{t-\alpha_{sw}}^{t} g_{sw}(t - \tau)d\tau = 9\left(1 - e^{-\alpha_{sw}}\right) \tag{3.3}$$

The schematic illustration for the switching action function, $F_{sw}(t)$ and critical switching action time, $F_{sw,c}$ is shown in Fig. 3.3. Figure 3.3 indicates that the moving direction is changed at the time $(t)_3$, when $F_{sw}(t)$ is equal to $F_{sw,c}$.

3.3   Further Application of the Model

Further application of the extended model is shown in the simulation for an evacuation from a building based on different conjectural scenarios. The focal interest of CBS-DE with switching action behavior model is to witness the performance of the switching action behavior model by demonstrating the change in destination during the evacuation process. In relation to that, a study of crowd evacuation was piloted at a local coastal airport, Langkawi International Airport (the LIA), located at Langkawi Island, Kedah, Malaysia. The airport was chosen as a study area due to its location facing the Andaman Sea.

The Langkawi Island is located in the northeastern of Peninsular Malaysia as shown in Fig. 3.4. Figure 3.5 shows the satellite view of the airport and its vicinity [2]. Two main buildings were considered as study areas which are the buildings of the airport itself and the nearby hotel.

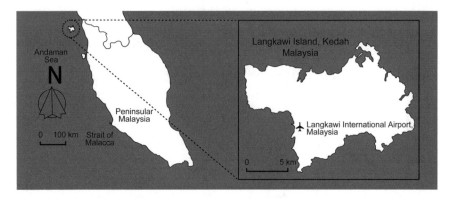

Fig. 3.4 The Langkawi Island, Malaysia

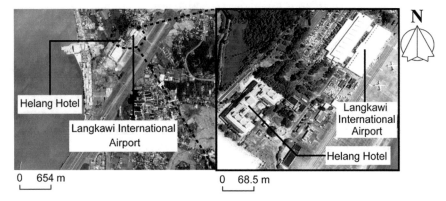

Fig. 3.5 The satellite view of the LIA and its vicinity. (*Source* Google Earth, 2014)

Figure 3.6 portrays the schematic flow of work in demonstrating the evacuation process at the LIA by considering the switching action behavior.

3.3.1 Data-Gathering

Data-gathering procedures like meetings, field observation and survey, and video footage were conducted to collect empirical data of the LIA and its vicinity (Fig. 3.7) Those procedures were purposely to obtain the exact layout of the LIA and its vicinity and to get an insight of the airport's daily activities including passengers' movement of "in" and "out" from the airport and an actual view of the situation inside the airport and the airport's vicinity.

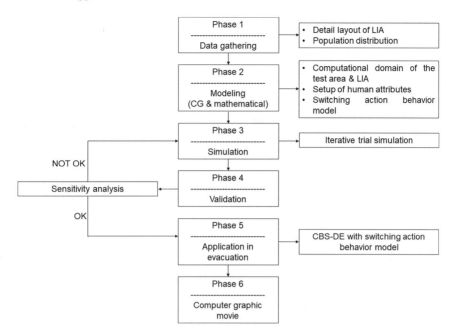

Fig. 3.6 The schematic flow of work

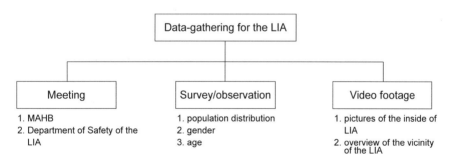

Fig. 3.7 Procedures of data-gathering

Data and information regarding the layout of the LIA were obtained from the meeting with the officer of the Malaysia Airport Holding Berhad (MAHB), Selangor, Malaysia. From that meeting, general layout of the airport and floor plan of the airport were requested for the purpose of this work. All materials requested were acquired in the form of hardcopies. The entire received drawings plans were redrawn by using AutoCAD® 2011 to have a digital format of drawings. Figure 3.8 illustrates the plan view of the Level 1 and Level 2 of the LIA.

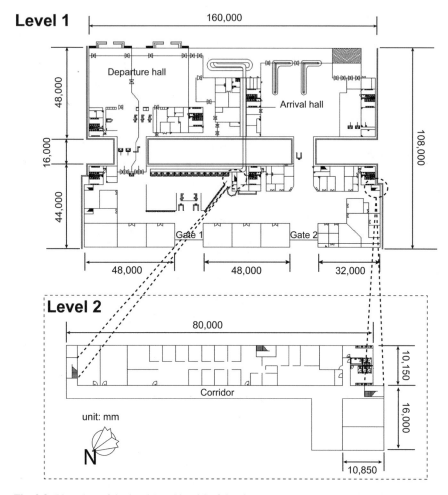

Fig. 3.8 Plan view of the level 1 and level 2 of the airport

Upon completion dealing with the officer of the MAHB, site-visit to the LIA was conducted. During seven days of the visit, meeting with the officer of the Department of Safety of the LIA, and observations/survey and video footage were performed to collect as much as possible useful, relevant and specific information and data at the LIA. Two types of data were looked-for in this work, namely data on the study area (the LIA and its vicinity) and data on human factors (population distribution, gender and age). Both data are crucial for modeling the computational domain and initializing the calculation.

Table 3.1 Survey form—population distribution

Date:	Time:	Location:		
Category	Gender	Age	Number of people	
			"In"	"Out"
1	Male	10–39		
2	Male	40–69		
3	Male	>70		
4	Female	10–39		
5	Female	40–69		
6	Female	>70		
7	Children	2–9		
Total				

During the meeting with the officer of the Department of Safety, the information of airport users was obtained based on passengers' movement of "in" and "out" in weekly and yearly schedule. In general, from 2007 to 2011 (5 years), the average passenger movements at LIA yearly was 1,283,392 persons (including local and international passenger). Based on recent data about an average of 5000 persons of airport users in a day who were "in" and "out" of the LIA. Meanwhile, the permanent staff who served at the airport (including ground staff, and safety officers) is 500 persons. Nevertheless, for the purpose of this contribution, the peak number of airport's user at specified time needs to be collected. Hence, through observation/survey, the distribution of population at the LIA was dictated manually. Four locations of placing observers were identified, which were at the entry/exit points of the LIA (Gate 1 and Gate 2), in front of arrival hall and departure hall's gate.

Initially, the airport's users (population) were identified subjectively based on seven different categories as described in the Table 3.1. Those categories are classified according to the age of the population; age of 10–39, 40–69, more than 70, for male and female, respectively and age of 2–9 for children. The identification of airport users was based on a subjective judgment of observers by looking at the individual appearance like apparels and characters. The numbers of "in" and "out" of the population were recorded manually by observers by filling up the survey form prepared as in Table 3.1. The data were collected for every one hour for three consecutive days. Three different times were selected based on the maximum number of flights landed and take-off within one hour. The times selected were from 9 am to 10 am, 11 am to 12 noon and 1 pm to 2 pm. Moreover, from the data of "in" and "out" collected, the maximum users in the airport during peak time are summarized in Table 3.2 as the population distribution. The ratio (the fifth column) refers to the ratio of each category to the total number of people observed. This ratio is crucial in setting person's attributes later.

Table 3.2 Population distribution in the airport during peak time

Category	Gender	Age	Number of people	Ratio [%]
1	Male	10–39	270	34.75
2	Male	40–69	125	16.09
3	Male	>70	0	0.00
4	Female	10–39	272	35.01
5	Female	40–69	92	11.84
6	Female	>70	0	0.00
7	Children	2–9	18	2.32
Total			777	100.00

Photo 3.1 Entrance of the LIA (Gate 2)

Finally, video recording was made to get an actual view of the inside and outside of the LIA. This footage was necessary in creating the calculation domain later whereby all the boundaries (walls) need to be defined before calculations are performed. Photos 3.1, 3.2, 3.3 and 3.4 show a few shots inside of the LIA and one photo for the view outside of the LIA.

Photo 3.2   The view from
the mezzanine floor

Photo 3.3   The check-in
area

Photo 3.4   The outside view
of the LIA

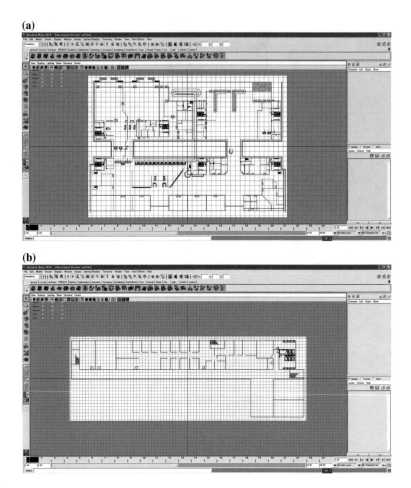

Fig. 3.9 **a** 2D image of the floor plan of the Level 1 of the LIA. **b** 2D image of the floor plan of the Level 2 of the LIA

3.3.2   Modeling the Computational Domain

In conjunction with creating the computational domain of the LIA, modeling a polygon mesh from a 2D reference image was imposed. Hence, the 2D images of the LIA were imported into Autodesk® MAYA® 2010 (MAYA) as a reference image for constructing the computational domain and 3D models. Figure 3.9 shows 2D images of the LIA imported into MAYA. Thereafter, modeling a polygon mesh from a reference image was conducted by using some of the polygon modeling tools by:

1. Working with components of a polygon mesh (faces, edges, and vertices)
2. Selecting faces, edges and vertices polygon meshes
3. Scaling and extruding faces on polygon meshes
4. Splitting vertices and subdividing polygon faces
5. Combining separate meshes into one meshes
6. Adding faces to an existing mesh
7. Using snap to grid.

Figure 3.10 shows the product of modeling a polygon mesh from 2D reference images for both levels 1 and 2 of the LIA. The product of the modeling was still in 2D models. As such, to build 3D models, the same procedures as above were applied together with the texturing and rendering processes. In MAYA, texturing is a process of modifying the appearance of the 3D models and rendering is a process of creating bitmap images of the model based on various shading, lighting and camera attributes. Figure 3.11a–c show some of the 3D scenes of the LIA created in MAYA.

Afterwards, the population distribution, boundaries, persons' attributes and domain's characteristics of the LIA are initialized on the.

3.3.3 Computational Domain and the HBS Tool

Figure 3.12 shows the schematic diagram of the computational domain of the LIA and its vicinity. The area covers approximately 109,242 m^2 with two main buildings: the airport (the LIA) and the nearby hotel, Helang Hotel. In this contribution, the investigation was related to the tsunami evacuation process. Hence, an area with a height more than 5.0 m is considered as an adequate level for evacuation area. This level is predicted from the previous history of tsunami height that hit Malaysia in December 2004. From the survey conducted, two locations of evacuation areas were identified; the first location was located on the Level 2 of the LIA (at the corridor) and the second location was located at the Level 2 of the hotel as an additional location. Both levels were 6.0 m in height, suitable as evacuation places.

On the other hand, the HBS tool was specially designed to enhance the capability of the simulator in specifying person's attributes and domain's characteristics. It is a scripting language which is embedded into MAYA acts as a plug-in component that adds a specific feature to the simulator. To make use of the HBS tool, Fig. 3.13 shows the flow chart of procedures imposed specifically in this contribution in defining a person's attributes and domain's characteristics. The details application of the HBS tool is elucidated in the Appendix A.

(a)

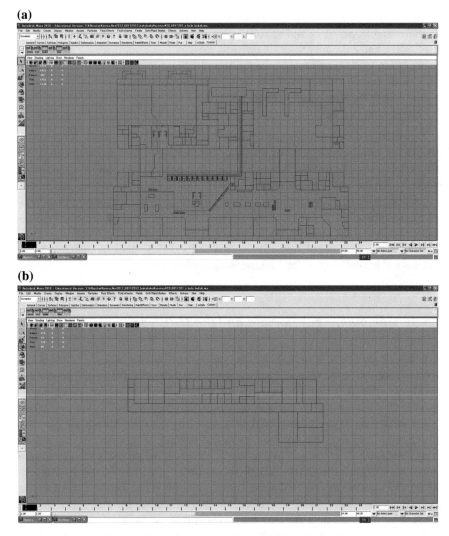

(b)

Fig. 3.10 a Modeling a polygon mesh from 2D reference images for the Level 1 of the LIA. **b** Modeling a polygon mesh from 2D reference images for the Level 2 of the LIA

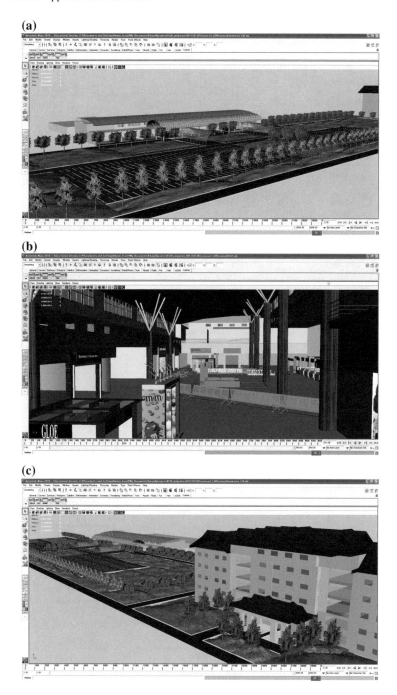

Fig. 3.11 a The 3D scene of the airport and its vicinity. **b** The 3D scene inside the LIA. **c** The 3D scene of the airport and its vicinity

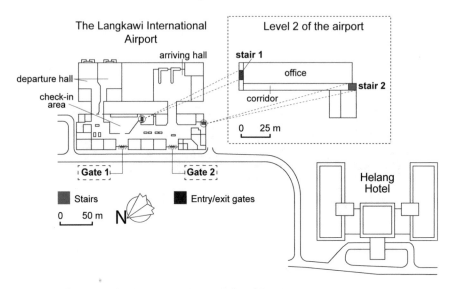

Fig. 3.12 The schematic diagram of the computational domain

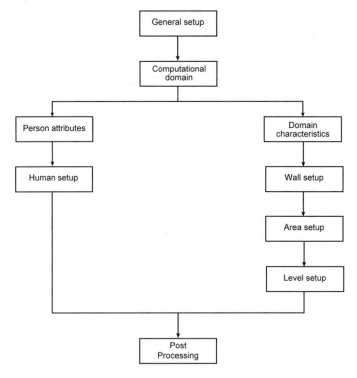

Fig. 3.13 Flow chart of procedure impose in defining person's attributes and domain's characteristics using the HBS tool

References

1. Department for Culture, Media and Sports: Guide to safety at sports ground, 5th edn. http://www.safetyatsportsgrounds.org.uk/publications/green-guide. Accessed 1 Aug 2014
2. Google Earth 7.1.3.22.3 version: Langkawi International Airport, Langkawi Island, Kedah, Malaysia. http://www.google.com/earth/index.html. Accessed 14 Aug 2014

Chapter 4
Application of DEM-Based Crowd Behavior Simulator with Switching Action Behavior Model

4.1 Calibration and Validation Settings

4.1.1 Simulation Domain and Condition

To examine the change in destination exhibited by the switching action behavior model, a few trial simulations were conducted on the proposed test computational domain as depicted in Fig. 4.1. The test computational domain consists of a corridor with a length of 21.0 m and a width of 3.0 m, connected with a rectangular evacuation area on the right-side and an initial position area of persons on the left-side. The designated evacuation area is 4.0 m in length and 3.0 m in width. The test computational domain is delimited by walls with thickness of 0.379 m. The capacity of the designated evacuation area is 35 persons, which gives the density of 2.9 persons/m^2. This number is determined based on the definition of the crowd density by Department for Culture, Media and Sports [1].

In the trial simulations, a total of 100 persons were considered and they were positioned randomly with no contacting condition between persons in the initial position area as shown in Fig. 4.1, with the initial velocities of zero. CBS-DE with switching action behavior model was employed. Persons were moving from the left-side of the test computational domain to the designated evacuation area on the right-side with an inflow rate of 1.9 person/s. The trial simulations associated with the parameter setting in the sensitivity analysis were described in the next subchapter.

© The Author(s), under exclusive license to Springer Nature Singapore Pte Ltd. 2019
N. Abd Rahman, *Crowd Behavior Simulation of Pedestrians During Evacuation Process*, SpringerBriefs in Computational Intelligence,
https://doi.org/10.1007/978-981-13-1846-7_4

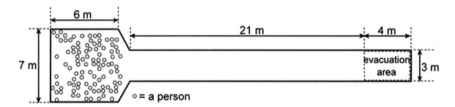

Fig. 4.1 Test computational domain and initial position of persons for trial simulation

4.1.2   Parameter Settings

In this contribution, the sensitivity analysis was performed to assess the impact of the model parameter for perception-response time, α_{sw} on the output of switching action behavior model. The validation was made with a focus on realistic motion of switching action behavior macroscopically. Further from the viewpoint of microscopic, the maximum contact force acting on each person was compared with the experimental results of comfortable loads for people observed by Smith and Lim [2].

On the basis of the above circumstances, a range of values of model parameter α_{sw}, which gives an appropriate switching action behavior visually, was calibrated first. This was achieved by conducting trial simulations on the test computational domain by employing CBS-DE with switching action behavior model. The calibrated range of values of model parameter α_{sw} in trial simulations that gives a realistic motion of switching action behavior also be checked for the safe level of contact forces.

Checking for the maximum contact force was performed within macroscopically calibrated range of values of model parameter. The maximum contact force to the recommended values of the parameter must satisfy the benchmark of the safe level of contact force.

4.1.3   Safe Level of Contact Force

The contact force acting on each person is a significant output from the current trial simulations. The measurement of potential for injury on a human had been reported previously. For instance, an experimental investigation on the level of comfortable loads for people had been conducted by [2]. Twenty-one people with the range of age from 20 to 25 years old had been tested on three barrier types loaded on upper and lower chest and abdomen. In their findings, the comfortable limits of loads were ranged from 175 N to 247 N. Hence, this range of the comfortable loads is used in this work as a benchmark for the safe level of contact forces for each person.

Table 4.1 The data for trial simulations

Trial simulation	Perception-response time, α_{sw} (s)	Critical switching action time factor, $F_{sw,c}$
Sim-1	0.100	0.856
Sim-2	0.125	1.058
Sim-3	0.150	1.254
Sim-4	0.175	1.445
Sim-5	0.200	1.631
Sim-6	0.225	1.813
Sim-7	0.250	1.991

4.2   Trial Simulations

4.2.1   Switching Action Behavior

The calibration is done through trial simulations for a change in direction by employing CBS-DE with switching action behavior model. In this work, a range of values for the parameter of perception-response time, α_{sw} was checked between 0.100 and 0.250 s with the interval of 0.025 s. In total there were seven calibrated values and hence seven trial simulations have been conducted (denoted as Sim-1–Sim-7). The data of the trial simulations, the range of values of the perception-response time α_{sw}, and the values of critical switching action time factor, $F_{sw,c}$ are summarized in Table 4.1. The critical switching action time factor, $F_{sw,c}$ was determined from Eq. 3.3.

Within the range of values for the parameter of perception-response time α_{sw}, performances of change in direction are visually realistic and acceptable. This are shown in the snapshots shown in Fig. 4.2a–g which illustrated the switching action behavior of persons who are changing their moving direction with different values of α_{sw} and their respective $F_{sw,c}$ for every 5 s in the time t ranging from 45 to 70 s. The circles in the snapshots represent persons and the arrows show the moving directions of persons. The white circles indicate persons who are moving from the initial position area to an evacuation area. Meanwhile, the black circles indicate persons who have changed their moving directions.

From the snapshots of all simulations (Sim-1–Sim-7), no significant difference can be seen in terms of the switching action behavior, realistic and reasonable switching action behavior can be shown. The person begin to change in his/her moving direction around the time $t = 42$ s (Sim-1 and Sim-2) and $t = 43$ s (Sim-3–Sim-7). The time necessary for completion of the switching action behavior is ranged from 21 to 24 s. The detailed results are concluded in Table 4.2. To investigate the results microscopically, the maximum contact force acting on each person was checked with a benchmark for the safe level of contact force for each person.

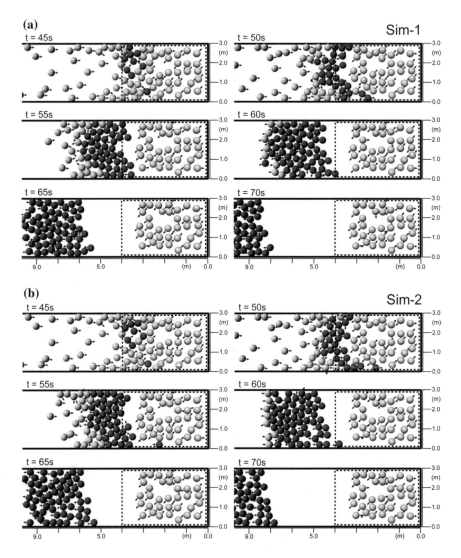

Fig. 4.2 **a** Snapshots of persons changing their moving direction ($\alpha_{sw} = 0.100$, $F_{sw,c} = 0.856$); **b** Snapshots of persons changing their moving direction ($\alpha_{sw} = 0.125$, $F_{sw,c} = 1.058$); **c** Snapshots of persons changing their moving direction ($\alpha_{sw} = 0.150$, $F_{sw,c} = 1.254$); **d** Snapshots of persons changing their moving direction ($\alpha_{sw} = 0.175$, $F_{sw,c} = 1.445$); **e** Snapshots of persons changing their moving direction ($\alpha_{sw} = 0.200$, $F_{sw,c} = 1.631$); **f** Snapshots of persons changing their moving direction ($\alpha_{sw} = 0.225$, $F_{sw,c} = 1.813$); **g** Snapshots of persons changing their moving direction ($\alpha_{sw} = 0.250$, $F_{sw,c} = 1.991$)

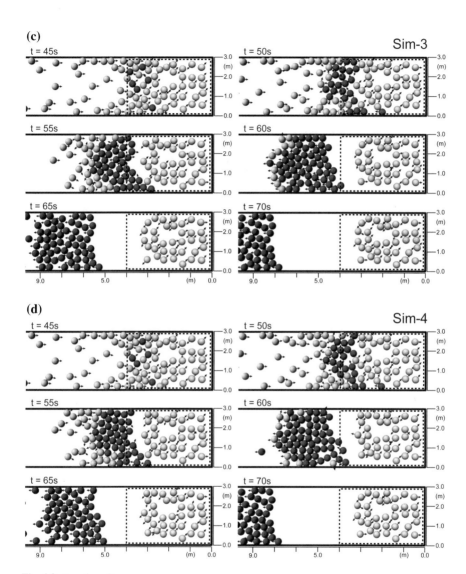

Fig. 4.2 (continued)

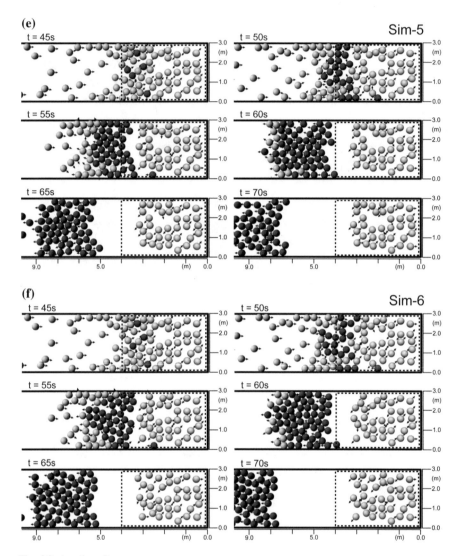

Fig. 4.2 (continued)

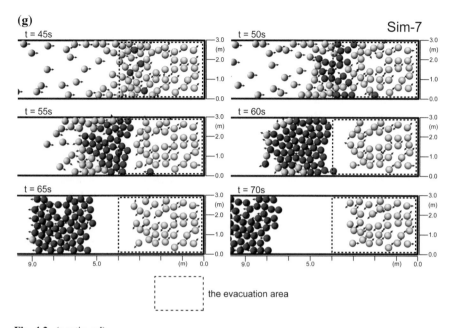

Fig. 4.2 (continued)

Table 4.2 The results obtained from trial simulations

Trial simulation	Evacuation completion time (s)	Time of first person to switch (s) (T_{sw1})	Time of switching action activity completed (s) (T_{sw2})	Duration of switching action (s) $(T_{sw2}) - (T_{sw1})$	Max. contact force, $F_{con.max.}$ (N)
Sim-1	97	42	66	24	725.73
Sim-2	97	42	65	23	208.63
Sim-3	96	43	65	22	376.98
Sim-4	97	43	64	21	166.44
Sim-5	97	43	64	21	447.43
Sim-6	97	43	65	22	563.89
Sim-7	98	43	66	23	309.42

4.2.2 Contact Force

The maximum contact force was a significant output in order to determine the value of the perception-response time α_{sw}. Hence, to ascertain the maximum contact force acting on each person is within the safe level of contact force, a time series of maximum contact force was plotted. Figure 4.3a–g show the time series of the maximum contact force acting on the person from the time $t = 40$–100 s for all the simulations (Sim-1–Sim-7). The vertical dashed-line in the graphs indicates the time range in which the switching action behavior is activated.

By considering the upper and lower limit of the comfortable loads in the range of 175–247 N, only $F_{sw,c} = 1.058$ (Sim-2) and 1.445 (Sim-4), with perception-response time $\alpha_{sw} = 0.125$ s (Sim-2) and 0.175 s (Sim-4), respectively, gave good results in terms of the safe level of contact force. The maximum value of the inter-element force with 208.63 N occurs at the time $t = 59$ s for the Sim-2 and at time $t = 53$ s for the Sim-4 with 166.44 N. Usually people tend to avoid contact with neighboring persons in the walking process. If we rely on this viewpoint, a parameter value in the Sim-4 should be chosen as the appropriate one rather than that in the Sim-2.

4.3 Evacuation Process Using CBS-DE with Switching Action Model

This subchapter shows the results of evacuation simulations using CBS-DE with switching action behavior model under two different conjectural scenarios. The purpose of these simulations is to reveal the effectiveness of the newly developed model for change in destination in the evacuation process. In conjunction with the evacuation process, a local coastal airport, named the Langkawi International Airport (LIA) which is situated at the Langkawi Island, Kedah, Malaysia, was used as a study area.

The appropriate scenarios were designed by investigating the evacuation process of 777 persons at the airport. All scenarios in this work are hypothetical and including switching action behavior that may happen in the evacuation process. The time necessary for completing the evacuation is recorded to show the effect of the switching action model.

4.3.1 Simulation Domain and Setup

In the current study, the simulation domain in the vicinity of the airport is illustrated in Fig. 4.4. The domain covers approximately 109,242 m^2 with two main buildings: the airport (LIA) and the nearby hotel. In this particular contribution, the evacuation investigation is related to the tsunami evacuation process, hence, an area with a height being more than 5.0 m is considered as an adequate level for evacuation area. From

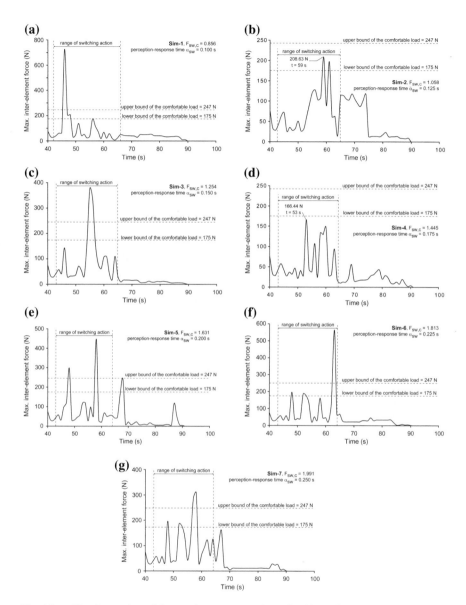

Fig. 4.3 a The time-series of the maximum contact force for Sim-1; **b** The time-series of the maximum contact force for Sim-2; **c** The time-series of the maximum contact force for Sim-3; **d** The time-series of the maximum contact force for Sim-4; **e** The time-series of the maximum contact force for Sim-5; **f** The time-series of the maximum contact force for Sim-6; **g** The time-series of the maximum contact force for Sim-7

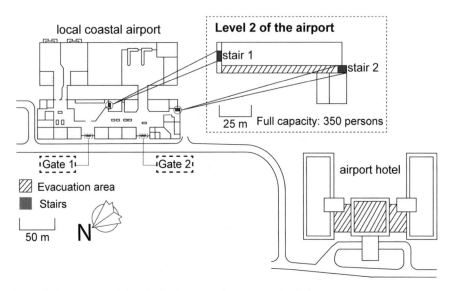

Fig. 4.4 Computational domain for the evacuation process simulation

the survey, two locations of evacuation areas are located; the first location is located on the Level 2 of the LIA and the second location is located at the Level 2 of the hotel as an additional location. Both evacuation areas are 6.0 m in height. Two main stairs, stair 1 and stair 2 as marked in Fig. 4.4 are used toward Level 2 of the airport, and only two main gates, namely Gate 1 and Gate 2 are available to evacuate to the hotel.

As for the simulation setup, human distribution in the airport was surveyed around the peak time as shown in Table 4.3. The results are categories into seven groups based on a range of age, gender, and the number of people. The average walking velocity for each group was taken from the analysis conducted by Abustan et al. [3] for Malaysian pedestrian. A total of 777 persons were observed during the survey, and all the 777 persons were arranged according to the on-site observation at the LIA. The initial positions of persons are as shown in Fig. 4.5.

4.3.2 Simulation of the Evacuation Process

Two conjectural scenarios of evacuation processes were considered in this contribution and named as Scene-1 and Scene-2, respectively. As recommended by Applied Technology Council [4], minimum square meter per occupant for a tsunami refuge is one square meter per person. Hence, on the basis of the available evacuation space at the Level 2 of the airport, only 350 persons can be accommodated. The balance of

Table 4.3 Detailed of human distribution at the airport during peak time

Category	Gender	Age	Number of people	Ratio (%)	Velocity (m/s)
1	Male	10–39	270	34.75	1.38
2	Male	40–69	125	16.09	1.14
3	Male	>70	0	0.00	0.99
4	Female	10–39	272	35.01	1.20
5	Female	40–69	92	11.84	1.04
6	Female	>70	0	0.00	0.89
7	Children	2–9	18	2.32	1.06
Total			777	100.00	

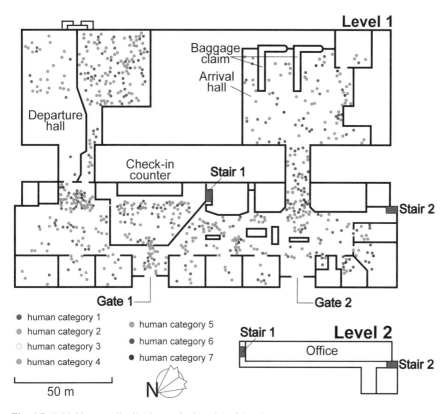

Fig. 4.5 Initial human distribution at the Level 1 of the airport

evacuees (427 persons) needs to change their direction and move to another evacuation area located at the Level 2 of the hotel. The evacuation routes for both scenarios are as depicted in Fig. 4.6 (Scene-1) and Fig. 4.7 (Scene-2), respectively.

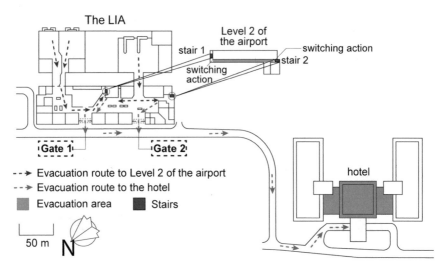

Fig. 4.6 Evacuation routes for the Scene-1

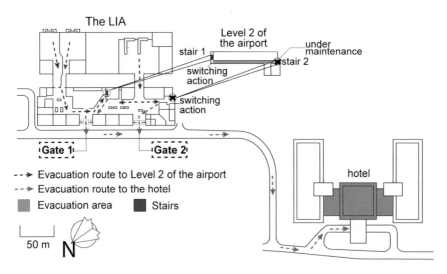

Fig. 4.7 Evacuation routes for the Scene-2

As for the Scene-1, initially, all the 777 persons are directly evacuate to the Level 2 of the airport using the stair 1 and stair 2. Once the capacity in the evacuation area achieved 350 persons, the balance of 427 persons has to change their directions toward another evacuation area set at the Level 2 of the hotel through the Gate 1 and

Gate 2. The time necessary for the completion of the evacuation was recorded under these conditions.

For the Scene-2, the following conditions are considered. The stair 2 of the airport is under maintenance as depicted in Fig. 4.7 and people do not notice about that problem in the beginning. In this Scene-2, initially people evacuate to the Level 2 of the airport in the same manner of the Scene-1. However, people who head to the stair 2 suddenly switch their directions towards the Level 2 of the airport hotel after they noticed the unavailable of stair 2. Concurrently, persons use the stair 1 evacuate to the evacuation area at the Level 2 of the LIA up to the limits. After the capacity in the evacuation area at the Level 2 of the airport is full with 350 persons, the balance persons change their directions and evacuate to the Level 2 of the hotel through the Gate 1 and Gate 2.

4.3.3   Simulation Outcomes

In the present simulations, CBS-DE with switching action behavior model in the evacuation process was well performed. Figure 4.8 shows the time series of the accumulative number of persons who completed the evacuation process for both scenes. In the Scene-1, the time to complete evacuation process is 1247 s, while in the Scene-2, the evacuation completion time is 1169 s. The difference of the evacuation completion time is 78 s (decrease by 6.25% in comparison with the Scene-1). Although there is only one route to the evacuation area in the airport in the Scene-2, the evacuation completion time of the Scene-2 decreases in comparison with that of the Scene-1 with two routes to the evacuation area in the airport. Therefore, it would be said that the evacuation plan including route restriction toward the evacuation area in the airport should be considered to make an effective plan.

In the present study, the focal interest is to show the ability of the switching action behavior model by demonstrating the change in destination in the evacuation process. The typical snapshots of switching action behavior at the Level 2 of the airport connected to stair 1 are shown in Fig. 4.9. In Fig. 4.9, after exceeding the capacity of the evacuation area, crowd behavior changing in their moving direction can be found. In the near future, further consideration should be given by comparison with additional observed phenomena to enhance the model in quantitative reproducibility.

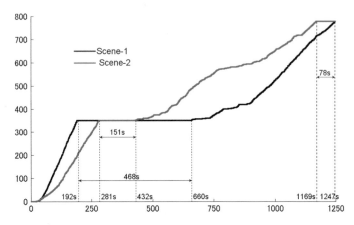

Fig. 4.8 Time series of the accumulative number of persons completed the evacuation for the Scene-1 and Scene-2

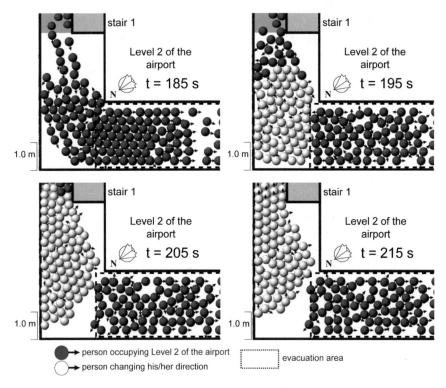

Fig. 4.9 Snapshots of switching action behavior (the Scene-1) at the Level 2 of the airport near the stair 1

References

1. Department for Culture, Media and Sports (2008) Guide to safety at sports ground, 5th edn. http://www.safetyatsportsgrounds.org.uk/publications/green-guide. Accessed 14 Aug 2014
2. Smith RA, Lim LB (1995) Experiments to investigate the level of 'comfortable' loads for people against crush barriers. Saf Sci 18(4):329–335. https://doi.org/10.1016/0925-7535(94)00052-5
3. Abustan MS, Harada E, Gotoh H (2012) Numerical simulation for evacuation process against tsunami disaster at Teluk Batik in Malaysia by multi-agent DEM model. Proc Coast Eng JSCE 3:56–60
4. Applied Technology Council (2008) Guidelines for design of structures for vertical evacuation from tsunamis. http://www.fema.gov/media-library/assets/documents/14708. Accessed 14 Aug 2014

Chapter 5
Recapitulation

5.1 Background

Many kinds of crowd evacuation models have been proposed and proved to be effective tools in evacuation decision making and risk evaluation. However, the existing evacuation models have not fully considered the uncertain behavior or phenomena of evacuees during the process of evacuation. Hence, there are certain areas need to be rectified and improved.

The overriding purpose of this book is to model a switching action behavior in a crowd that has a significant contribution to the study of a crowd behavior generally and in an evacuation process specifically. To achieve the objectives of this study, the existing simulator, Crowd Behavior Simulator for Disaster Evacuation (CBS-DE) based on the Distinct Element Method (DEM) has been utilized and extended accordingly. The switching action behavior model has been incorporated into CBS-DE and validated. Subsequently, the newly developed model has been demonstrated in evacuation simulation which was conducted at the Langkawi International Airport, Langkawi Island, Malaysia, as a research background. 3D computer graphic movies were utilized to visualize the simulated evacuation processes. The effects of the developed models were shown in the context of realistic motion of the crowd in a visual basis, evacuation completion time, and partially comparing with empirical data available.

Overall, the main objectives are achieved accordingly. Nevertheless, there are areas for improvements identified. Furthermore, some limitations in this research need to be tackled appropriately for future research.

© The Author(s), under exclusive license to Springer Nature Singapore Pte Ltd. 2019
N. Abd Rahman, *Crowd Behavior Simulation of Pedestrians During Evacuation Process*, SpringerBriefs in Computational Intelligence,
https://doi.org/10.1007/978-981-13-1846-7_5

5.2 Recapitulation

The idea to model and simulate switching action behavior in a crowd during the evacuation process was triggered after observations conducted at the Langkawi International Airport, (the LIA), Malaysia. From the observations, in the airport itself, there is no appropriate evacuation area that can be utilized to accommodate the crowd. The only available space as an evacuation area is on the Level 2 of the LIA, which is 6.0 m in height and available space of 350 m^2. As recommended by Applied Technology Council [1], minimum square meter per occupant for a tsunami refuge is one square meter per person. Hence, based on the available evacuation space at the Level 2 of the airport, only 350 persons can be accommodated.

With regards to the aforementioned situation, switching action behavior has been studied to simulate a scenario in which the surplus evacuees are required to change their destination from the Level 2 of the airport toward another evacuation area in the evacuation process. For that purpose, in this contribution, switching action behavior has been postulated based on the motion of other persons in the perception domain. Vision is the main source of information to control persons' motion. Hence, switching action behavior of each evacuee is determined according to the motion of other evacuees in his/her perception domain in the defined switching action time. The switching action time is formulated from the proposed switching action function, $F_{sw}(t)$. To avoid an artifact flow of switching action behavior, the switching action function, $F_{sw}(t)$ is only triggered if the number density of evacuees in the perception domain is more than 9, and the average walking velocity of these evacuees is less than 1.0 m/s.

To describe the switching action function, $F_{sw}(t)$, a convolution integral of input and unit response functions has been employed since it is the mathematical representation of the physical process with response. The exponentially-decaying function has been imposed as the unit response function which can contribute a computational stability. Concurrently, the critical switching action time, $F_{sw,c}$ has been introduced and calculated by considering Eqs. (3.1) and (3.2). $F_{sw,c}$ varies according to the congestion and perception-response time α_{sw}. At this point, switching action is only occurred if the switching action time, $F_{sw}(t)$ exceeds the critical switching action time, $F_{sw,c}$.

Due to the absence of empirical data in switching action behavior, the sensitivity analysis has been performed in order to calibrate the model parameter (perception-response time α_{sw}). Iterative trial simulations were conducted within prescribed model parameter values, which is from 0.100 to 0.250 with the increment of 0.025. The primary attention is to obtain a realistic motion of switching action behavior visually. To enhance the realism and valid simulation results, the maximum contact force acting on each evacuee has been monitored so that it is not exceeding the stipulated safe level of contact force which is in the range of 175 N to 247 N.

Based on the trial simulations conducted, the motion of switching action can be produced. Visually, the motions are quite realistic. This was shown from the trial simulation outcomes. Nevertheless, the reliability of the model still needs to be enhanced in the future. The absence of empirical data currently can be overcome by conducting controlled switching action behavior experiments. By acquiring such experimental data, the model can be validated by comparing the results with the experimental data obtained.

Reference

1. Applied Technology Council (2008) Guidelines for design of structures for vertical evacuation from tsunamis. http://www.fema.gov/media-library/assets/documents/14708. Accessed 14 Aug 2014

Appendix
Human Behavior Simulator (HBS) Tool

A.1 The Details Application of HBS Tool

From the Script Editor in MAYA, the HBS text file is loaded and executed.

1. To open the Script Editor, as shown in Fig. A.1a, select: Window > General Editors > Script Editor.
2. The Script Editor box will appear immediately, as in Fig. A.1b.
3. In the bottom part of the box, the HBS text file is loaded, saved and executed.
4. To ease the modeling work in using the tool, the HBS text file in the Script Editor box is dragged by using the mouse onto the Custom Shelf. Figure A.2 shows the HBS tool icon on the Custom Shelf.
5. By clicking the HBS tool icon, the HBS box will pop-up and by default the General menu interface is displayed as shown in Fig. A.3.

At the top of the tool box (Fig. A.3), it can be viewed a Menu bar consisting of eight menus:

1	General	5	Area
2	CV tools	6	Level
3	Human	7	Calculation
4	Wall	8	PostProcessor

A.1.1 General Menu

The General menu is where the general setup is made before initializing modeling works. From the General menu, basic requirements are setup as follows:

Fig. A.1 HBS scripting in
MAYA. **a** To open Script
Editor application; **b** Script
Editor box

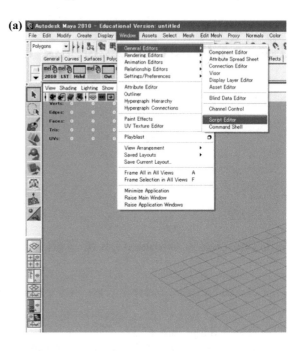

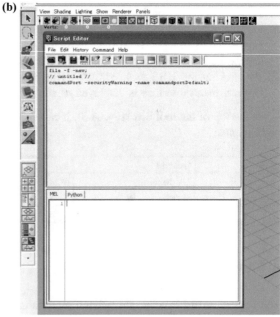

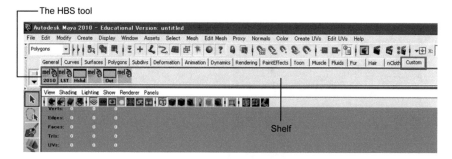

Fig. A.2 The HBS tool

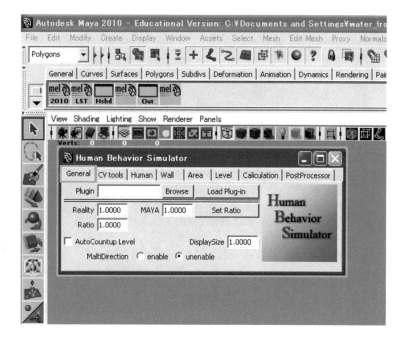

Fig. A.3 The HBS box—General menu

1. Activate the tool—the tool needs to be activated by bringing in the MLL file extension.
2. From the Plugin box in the General menu (Fig. A.4a), the Browse tap is clicked and the HBS.mll file is selected from the folder it is saved.
3. Then the Load Plug-in tap is clicked and automatically the HBS tool is initiated (Fig. A.4b).
4. Set the scale ratio—the ratio of the object scale between the on-site research background (Reality) and in MAYA environment (MAYA) needs to be set up.
5. The ratio between the reality and MAYA is inserted in the Ratio box. Afterwards the Set Ratio tap is clicked.

(a)
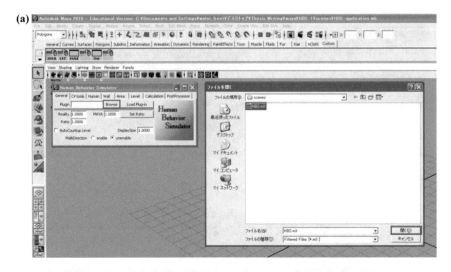

(b)
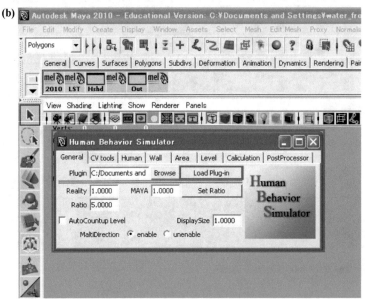

Fig. A.4 a To load the HBS.mll file; **b** The HBS.mll file is loaded

6. Establish the number of walking directions—on the MultiDirection, the enable or unenable box need to be selected. The MultiDirection refers to the pedestrian movement in a crowd either one direction (unenable) or more than one direction (enable).
7. Other requirements—the items of AutoCountup Level and DisplaySize, are set unchecked and 1.0000, respectively.

A.1.2 CV Tools Menu

The menu of CV tools as shown in Fig. A.5 is a tool that allows us to examine and edit coordinates of a vertex in MAYA.

1. To examine the coordinates of a vertex, the vertex of a polygon mesh in MAYA is selected, and from the CV tools, the appropriate direction of coordinate, like the X, and/or Y, and/or Z are checked and the Refer CV tap is pressed.
2. To edit the coordinates of a vertex, again the appropriate direction of coordinate is checked and the edited value of the coordinates are inserted in the X =, and/or Y =, and/or Z = box(s), and the Move CV tap is pressed. The Relative item is left unchecked.

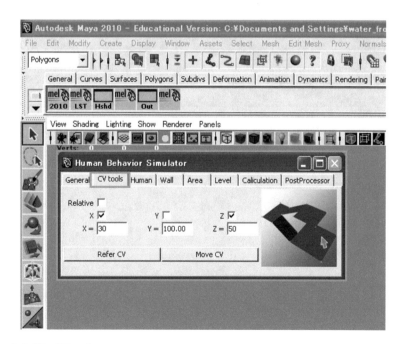

Fig. A.5 The CV tools menu

A.1.3 Human Menu

The Human menu (Fig. A.6) is a menu where the setup of person's attributes is generated before a simulation is performed and only initialized after the computational domain is ready. Basic person's attributes are set as follows:

1. Level in this menu refers to the numbering of horizontal plane ('faces' in MAYA environment) in which a person is situated/ positioned, with regard to its elevation. In the Level box, the numbering of level for the position of person is set 1, 2, 3 and so on. If there are no other levels in the domain, the Level is simply set as 1.
2. The number of persons and initial position—the total number of persons on faces in the domain is inserted in the N of Human box (refer to Fig. A.7a).
3. To load the number of persons on the domain, the faces on the domain are selected and the Output InitHuman tap is clicked. Automatically initial positions of persons with the prescribed numbers are displayed randomly (Fig. A.7b). Persons are represented by spheres.
4. To delete persons appeared on the computational domain, the Delete InitHuman tap is clicked and to reload again persons on the computational domain, the Load InitHuman tap is clicked.

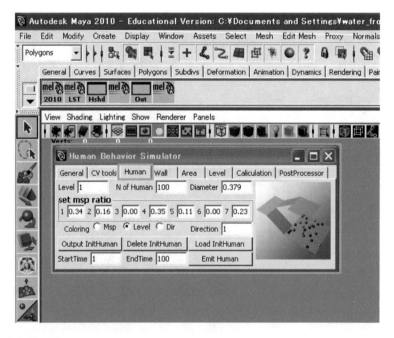

Fig. A.6 The Human menu

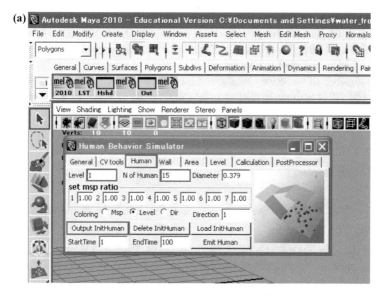

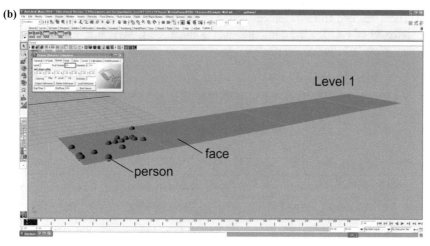

Fig. A.7 **a** The setting of number of persons; **b** The position of person on the domain with the prescribed numbers

5. The diameter of persons—the diameter of person is set homogeneously as 0.379 m in this research, hence in the Diameter box, 0.379 is inserted.
6. The msp ratio—the msp ratio refers to the ratio of each person's category in the population distribution (as calculated and showed in Table 5.2). In each box of category, the value of ratio is inserted.

7. The coloring—different colors are imposed on persons in relation to either category of persons (Msp) or position of persons (Level) or movement direction of persons (Dir). The choice of attributes to be colored is simply by selecting any of Msp or Level or Dir items.
8. The number of direction—the Direction in this menu refers to the direction of persons. This is directly related to MultiDirection selected in the General menu earlier.

A.1.4 Wall Menu

From the Wall menu, as shown in Fig. A.8, the level and diameter of the walls' boundary are established.

1. Level item in this menu is similar to Level item defined in Human menu, but this time refers to the walls' level.
2. The diameter in this menu refers to the wall's diameter and set as equal to human's diameter, 0.379 m.
3. There are two rows of taps available at the bottom of the Wall menu. For the purpose of creating the computational domain, only the first row taps are used.

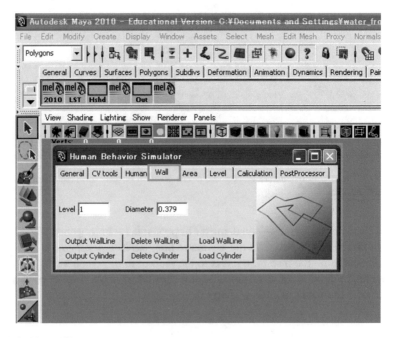

Fig. A.8 The Wall menu

4. The functions of those taps are similar to those in Human menu taps (Output InitHuman, Delete InitHuman and Load InitHuman).
5. The Output WallLine tap is to display the walls defined, the Delete WallLine tap is to delete the defined walls and the Load WallLine tap is to redisplay the walls defined earlier.

In defining the wall line, the curve lines on the domain need to be created.

1. From the Create menu, select the CV Curve Tool box on the right (Fig. A.9a).
2. In the CV Curve Settings, the Curve degree of 1 Linear is set (Fig. A.9b).
3. Afterwards, by using a mouse, curve is created on the domain by clicking on the vertex of the polygon mesh that defines the wall boundaries of the domain.
4. To display the wall line, the curve created is selected and the Output Walline tap is pressed. Figure A.9c shows the wall line on the domain.

A.1.5 Area Menu

In the Area menu (Fig. A.10), the directions of persons on faces are established.

1. The curves with 1 linear curve degree in MAYA are drawn on a polygon mesh by using the CV Curve Tool.
2. At this stage the Level box is set by 1 and the Goal box is set by 0.
3. Select curves and faces of curves, then Output Area tap is clicked (Fig. A.11a).
4. The directions represented with cones will be displayed on the domain (Fig. A.11b).
5. To delete the displayed arrow on the domain, the Delete Area tap is clicked and to reload the arrow on the domain, the Load Area tap is pressed.
6. The Goal is an item to set the final destination of persons which reflects the evacuation area.
7. To set the goal in a calculation, as shown in Fig. A.11c, in the Goal box, number 1 is inserted.
8. This number is related to the Direction set earlier in the General menu.
9. In the Direction Number box, number 1 is put in. This value is in relation to the number of Goal set earlier.
10. Afterwards, the cone where the goal wants to be set is selected and at the bottom part of the menu, the Goal item is selected and finally, the Edit Area is pressed.
11. The cone selected will change the direction upwards as displayed in Fig. A.11d.

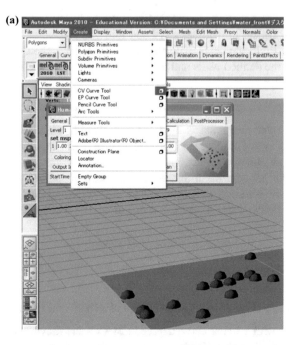

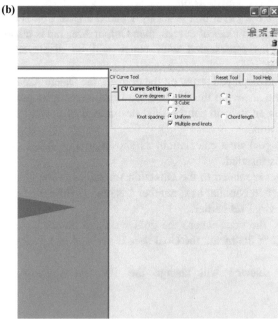

Fig. A.9 a The CV Curve Tool box on the right is selected; **b** The Curve degree of 1 Linear is set;
c The Output Walline tap is selected

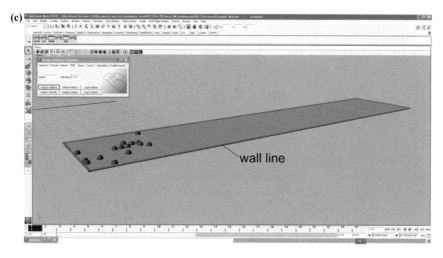

Fig. A.9 (continued)

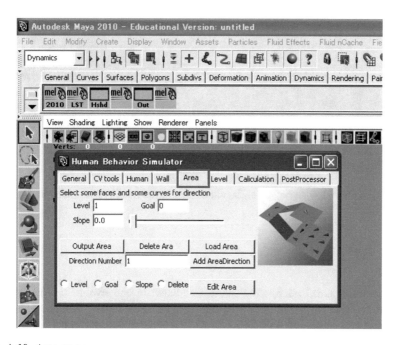

Fig. A.10 Area menu

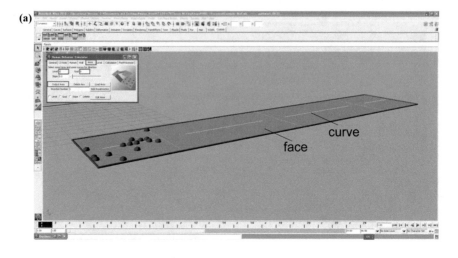

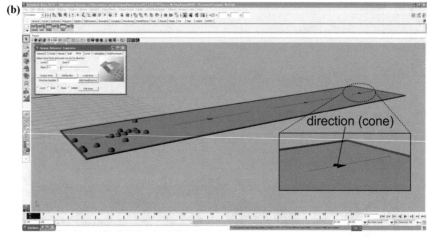

Fig. A.11 a The direction setting; **b** The direction of the area; **c** The goal setting; **d** The goal

A.1.6 Level Menu

Level menu is used to define the connection between one level to another level. If there is no other levels, in the Level and Connect Level boxes, number 1 and 2 is inserted, respectively, as shown in Fig. A.12. Then the Output ConnectLevel tap is clicked.

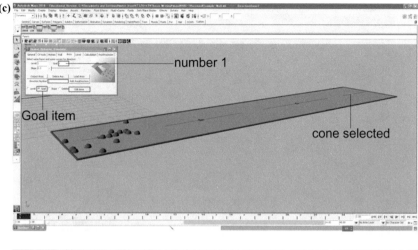

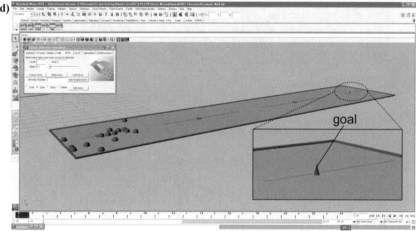

Fig. A.11 (continued)

A.1.7 PostProcessor Menu

After defining person's attributes and domain's characteristics, the simulation calculation is performed. After the calculation is completed, to visualize the motion of persons on the domain, the PostProcessor menu is selected (see Fig. A.13). The boxes of the StartFrame and EndFrame are related to the simulation calculation time. To visualize the motion, the Read File tap is selected.

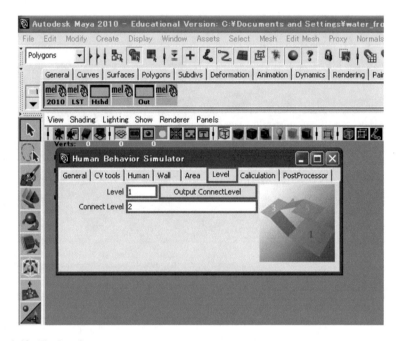

Fig. A.12 The Level menu

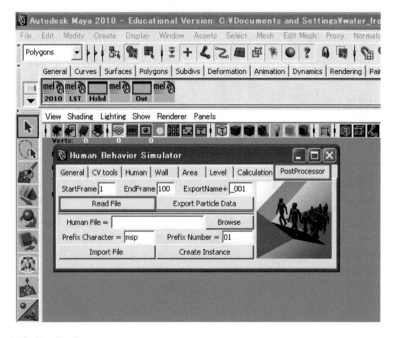

Fig. A.13 The PostProcessor menu

Printed in the United States
By Bookmasters